Nature That Makes Us Human

Nature That Makes Us Human

*Why We Keep Destroying Nature and
How We Can Stop Doing So*

Michel Loreau

OXFORD
UNIVERSITY PRESS

Oxford University Press is a department of the University of Oxford. It furthers
the University's objective of excellence in research, scholarship, and education
by publishing worldwide. Oxford is a registered trade mark of Oxford University
Press in the UK and certain other countries.

Published in the United States of America by Oxford University Press
198 Madison Avenue, New York, NY 10016, United States of America.

Library of Congress Cataloging-in-Publication Data
Names: Loreau, Michel, author.
Title: Nature that makes us human : why we keep destroying nature and how
we can stop doing so / Michel Loreau.
Description: New York : Oxford University Press, [2023] |
Includes bibliographical references and index.
Identifiers: LCCN 2022044802 (print) | LCCN 2022044803 (ebook) |
ISBN 9780197628430 (Hardback) | ISBN 9780197628454 (epub)
Subjects: LCSH: Nature—Effect of human beings on. | Environmentalism.
Classification: LCC GF75.L64 2023 (print) | LCC GF75 (ebook) |
DDC 304.2/8—dc23/eng20230111
LC record available at https://lccn.loc.gov/2022044802
LC ebook record available at https://lccn.loc.gov/2022044803

DOI: 10.1093/oso/9780197628430.001.0001

Printed by Sheridan Books, Inc., United States of America

Contents

Acknowledgments

I would like to extend my heartfelt thanks to all those who helped me in the writing of this book: first, my wife Claire de Mazancourt, who supported and encouraged me in this endeavor from start to finish; second, my collaborator Gladys Barragan-Jason, who showed fantastic enthusiasm for my book and provided me with invaluable comments; and third, my sister Dominique Loreau and my former collaborators Kirsten Henderson and Dalila Booth, who helped me to improve the book's content and form.

Introduction

Climate change, biodiversity loss, pollution, depletion of resources, new emerging diseases: year after year, new scientific reports are raising the alarm about the disastrous ecological and societal consequences, present or future, of the unbridled development of human activities. Almost everyone today is either clearly aware, or at least has a vague perception, that humanity is heading for major natural upheavals that threaten the very existence of contemporary human societies. A growing number of people are expressing their concerns about this state of affairs, or even are engaging in small-scale transformation of their lifestyles. And yet, nothing—or very little—is being done collectively to stop or slow down the social and economic machine launched at breakneck speed toward the wall of our planet's ecological limits, which is getting dangerously close. Scientists continue to sound the alarm, politicians get busy, international conferences are taking place, people are worried, but nearly everything continues as before. Only the recent Covid-19 pandemic has shaken humanity out of its apparent lethargy: suddenly the threat was perceived as immediate and—unthinkable until then—more than half of humanity agreed to remain confined for several months, thereby sharply reducing its economic activity and its ecological impact. But no sooner did the pandemic appear to be slowing down than powerful voices called for a resumption of economic activity, when all the evidence suggests that the damage from climate change and biodiversity loss will soon be far greater than that of the Covid-19 pandemic.

As a scientific ecologist, I have devoted most of my research activities to establishing, in a rigorous and systematic way, the consequences of current biodiversity loss for the functioning and stability of ecosystems and its longer-term consequences for human societies (Loreau et al. 2022). I have also sought to raise awareness among the general public and political decision-makers at the highest level of the importance of biodiversity loss and its ecological and societal consequences. I have devoted a lot of time and energy to promoting, on an international level, an integrative biodiversity science, as well as a science-policy interface in the field of biodiversity and ecosystems (Loreau 2010). These efforts have resulted, among others, in the creation of the IPBES

Nature That Makes Us Human. Michel Loreau, Oxford University Press. © Oxford University Press 2023.
DOI: 10.1093/oso/9780197628430.003.0001

(Intergovernmental science-policy Platform on Biodiversity and Ecosystem Services). I believe that all these efforts have been useful, but, like many of my fellow scientists, I have come to recognize that knowledge is not enough to generate action. The high level of understanding that science has reached about climate change, biodiversity loss, and their consequences is now more than sufficient to justify a profound transformation of our societies, our way of life, and our relationship with nature before it is too late.

Why, then, do we collectively continue to destroy nature and let the climate change when science tells us clearly that, in doing so, we are in danger of running to our own collective destruction? There are a number of reasons for this deplorable state of affairs. On an individual level, it is difficult both to project oneself into the distant future and to give up the comforts of modern life. On a collective scale, it is difficult to agree on a fair distribution of the efforts to be made. In general, it is simply difficult to change unless we are forced to. But there are also deeper reasons for this, which are less immediately apparent as they permeate contemporary thinking. In particular, the separation between humans and nature is one of the most powerful myths of Western civilization, a myth deeply rooted in the great monotheistic religions and in modernity. The protection of nature clashes with the collective belief that humans have the right, and even the duty, to dominate nature and transform it for their own benefit.

Many authors, writers, and philosophers have already written about the causes and consequences of the separation between humans and nature in modern society. Reading the many books on this topic, however, it seemed to me that something was missing. First, each author naturally tends to develop an idea that is dear to him or her and thus to focus on a particular aspect of the general problem. This view from a particular perspective is often very rich and intellectually stimulating, but at the same time it does not allow the problem to be considered in its entirety and to draw all its consequences. In particular, as an ecologist, I felt that many of these contributions lacked a biological dimension, which is fundamental for understanding both humans and nature. Second, many of the books on this topic are essentially critical, that is, they question a number of presuppositions or historical developments that have led to the separation between humans and nature as we know it today. They do not, however, seek to lay the foundations for an alternative worldview that would enable us to overcome the global ecological crisis that contemporary society is entering head-on.

This book is the result of my efforts to fill these gaps. I have used knowledge from a variety of disciplines and approaches—including biology, ecology, physics, psychology, anthropology, economics, history, philosophy, and personal development—to try to understand why we keep destroying nature today and how we could stop destroying it tomorrow. I realize that this is an ambitious goal, that my knowledge is limited, and thus that my book might disappoint some specialists in the various disciplines from which I use certain elements to feed my argument. In particular, several chapters of my book give pride of place to philosophy, and I am not a philosopher. Other chapters deal with economics, and I am no more of an economist. But, as the saying goes, economics is too serious a thing to be left in the hands of economists. I likewise believe that philosophy is too serious a thing to be left in the hands of philosophers, because it touches on the worldview that guides all our thoughts and actions. Anyone who is interested in the meaning of his or her life and in his or her place in the world should be able to call upon and use philosophy. The same is true, by the way, of my own scientific discipline, ecology. Ecology has been used by many people for all kinds of purposes for the last sixty years or so. Personally, I do not see this as a problem as long as everyone remains aware of the limits of whatever use they make of it, according to their skills and knowledge.

Recent scientific findings have also given me the firm conviction that we are going to face profound ecological and social upheavals in the coming decades, and thus that we can no longer afford to continue thinking as we have in the past. The so-called natural and social worlds will inevitably become increasingly intertwined, so that the traditional division of the scientific endeavor into "natural" and "human" sciences no longer provides us with the means to understand current challenges. More fundamentally, I will show in this book that the separation between body and mind underlies the modern separation between humans and nature and that a reunification of the entirety of human knowledge, whether it comes from the body or the mind, is essential to recover the lost unity of humans and nature. Therefore, I believe that we no longer have a choice: we need to return to a more holistic, integrative, and universal approach to human knowledge if we are to have any chance of overcoming the ecological and societal crisis that lies ahead.

This book contains two parts. Part I aims to deconstruct the myths of modern society that generate and perpetuate human domination over nature. It begins by establishing some biological foundations of human nature, which most of the ideologues of modernity have deliberately denied or ignored to justify the superiority of the human species over the rest of the living world. It then summarizes the main causes and historical stages that have led to the divorce between humans and nature as we know it today. Finally, it seeks to unpack the main founding myths of modernity that still shape our way of thinking and that lead us to accept the subjugation and destruction of nature.

These powerful myths include the duality of subject and object, the duality of matter and spirit, the rationality of modern economy, and the centrality of humans in the modern worldview.

My aim in Part I of the book is not to make an exhaustive critique of modern rationalism, which has already been done by numerous authors from multiple angles. Rather, the question that concerns me is the following. Modern rationalism is a fairly recent ideological construction, although in truth, as we shall see, it is the result of a long historical trajectory that has unfolded since the Neolithic revolution. Science, which is one of its most emblematic products, is constantly accumulating knowledge that calls into question its very foundations and shows that it is only one worldview among many others. It also demonstrates unequivocally that modern society is heading for its own demise by endangering the biosphere and the climate system that allowed it to flourish. Why, then, despite the repeated questioning and warnings of science, does the belief in modern rationalism remain so tenacious? Why is it that the critique of this worldview, which has already been made on numerous occasions, remains largely inaudible outside a relatively small circle of philosophers or believers? Why do the very serious threats posed by climate change, biodiversity loss, and changes in the functioning of the biosphere for present and future human societies almost systematically take a back seat in political decisions, or are simply denied? In order to answer these questions, it is essential to clearly identify the core collective beliefs that lie at the root of modern society's inability to substantially modify its relationship with nature. These beliefs are a powerful obstacle, without us even being aware of it, to all individual or collective attempts to overcome the ecological crisis we are entering on a planetary scale. This obstacle must be removed so that a new worldview more suited to current conditions can emerge.

In contrast, Part II of the book seeks to identify a few avenues that could enable human societies to break the current deadlock and take a new path, that of the flourishing of life on Earth. This path is based on a simple observation: humans have a nature that defines them as a unique species beyond their many cultural differences, and this nature is not only made up of flesh and bone, but also of a set of fundamental human needs. These needs are more than the basic physiological needs that are usually discussed; they also define the deep aspirations that all human beings share. The expression and satisfaction of their fundamental needs reconnects people to nature, as these needs are the manifestation of life within them. At the same time, it restores the unity of body and mind and thus of the different forms of knowledge that come from body and mind. For as long as we ignore the body as the primary source of knowledge, we cannot prevent our mind from reasserting its supremacy over the body, and, as a result, the supremacy of humans over nature. The economy, which today is essentially concerned with the creation, accumulation, and distribution of abstract social wealth, must place itself at the service of life, and, in particular, of the satisfaction of fundamental human needs. Only in this way can the current conflict between economic development, human development, and nature conservation be resolved.

This book will undoubtedly leave many questions unanswered, and that is fine. It does not pretend to provide a set of ready-made answers to questions that are among the most fundamental that humanity has asked itself since the Neolithic revolution. It is up to present generations to invent a new relationship with nature that will allow the human species to flourish in the midst of a flourishing nature. This challenge, unprecedented in history, requires humans to rethink almost everything they have been used to, from their existential aspirations to the form and content of the contemporary global economy. A book cannot claim to meet such a formidable challenge on its own. I just hope that my book will shed some useful light on the questions that need to be asked and how they can be answered.

PART I
HUMANS VERSUS NATURE

1

Homo sapiens, a species among many others . . . but not quite like the others

Popular imagery has it that the Renaissance and the Enlightenment freed humans from the backward thinking of the Middle Ages, a dark age in human history. Nothing could be further from the truth. In particular, as we will see in the next chapter, the modern philosophical thought that emerged from the Renaissance was largely based on the worldview developed by Christianity and the other monotheistic religions. In this worldview, humans possess characteristics that make them a species apart, radically different from any other living species. Humans were created in the image of God, the Bible tells us, with the mission of conquering and dominating the Earth and all the other living beings on it. Humans have a soul, Descartes echoes, unlike plants and animals, which can therefore be considered as machines at our disposal. This concept may make people smile today, but it is nevertheless the one that still permeates all contemporary civilization. The great philosopher Martin Heidegger said less than a century ago that man is a "world creator," whereas animals are "world impoverished." And despite the immense progress in scientific knowledge since Descartes, contemporary science continues to be fascinated by what makes humans different from all other species.

Before examining the origins and consequences of the separation between humans and nature that has been increasingly asserted in the course of the history of Western thought, let us begin by establishing some scientific foundations on the point of convergence of humans and nature, namely the nature of humans themselves. Are humans a unique species, radically different from any other living species, or, on the contrary, are they an animal species like any other?

First of all, it should be noted that the question thus posed is ambiguous, which has led to endless debates between philosophers and biologists. From a biological point of view, every species is by definition unique, in the sense that it has a set of characteristics that distinguish it from other species. At the same time, every species is connected to the same genealogical tree of life; it has a history and certain characteristics common to all other species. Humans

All figures courtesy of the author

Nature That Makes Us Human. Michel Loreau, Oxford University Press. © Oxford University Press 2023.
DOI: 10.1093/oso/9780197628430.003.0002

are like the humpback whale, the housefly, or the *E. coli* bacteria in our gut: we share a wide range of structures and processes that ensure the basic functioning and reproduction of our cells. Humans are therefore necessarily both a unique species and a species like any other, just as the humpback whale, the housefly, and *E. coli*.

Once this is understood, the debate on human nature suddenly looks an awful lot like the old question of whether a bottle is half empty or half full. But in the case of the human nature debate, this comparison obscures the high stakes of the different worldviews implied by the half-empty and half-full bottles. For example, it is not difficult to see that the view that humans differ radically from all other species (the half-empty bottle) leads quite naturally to the thesis of the separation between humans and nature. Conversely, if humans are essentially an animal species like any other (the half-full bottle), there is little reason to consider that they are not an integral part of nature.

From a strictly biological point of view, the question has lost much of its relevance today. Scientific advances in recent decades, notably in molecular biology, neurosciences, and animal psychology, have shown that the bottle is not half full, but over 99% full. Humans are so similar to their primate cousins that the idea of a radical break between humans and all other animal species seems almost absurd. For example, the genetic material of the human species differs from that of the chimpanzee by only about 1% (the figures differ somewhat depending on the method used) and most of these differences are in so-called neutral genes, i.e., genes that have no obvious effect on the characteristics of the two species. We are therefore left to speculate whether a difference of the order of one-tenth of a percent in genetic material could take humans out of the animal kingdom.

However implausible it may seem, this possibility should not be completely ruled out, though, since it is conceivable that a major innovation involving a small number of genes could have occurred recently in the evolution of the human species. We must therefore seriously examine the arguments put forward by those who still believe in a radical break between humans and the rest of the tree of life.

The first thing to note is that all these arguments refer to the superior intellectual capacities of the human species (Schweitzer & Notarbartolo-di-Sciara 2009). The list of properties presented as unique to humans is long; it includes, in particular, their soul, self-awareness, empathy, thought, language, culture, morality, and use of tools—all characteristics that highlight the intelligence of humans as opposed to "beasts." The word "beast" itself comes from the old French *beste*, which also means "stupid," "dumb." Plants and animals are supposed to be stupid; by contrast, humans are supposed to be intelligent. Even assuming that the above properties were really unique to humans (which, as we shall see, they are not), they highlight the circular nature of the reasoning based on these arguments: first, a list is drawn up of what is supposed to be unique to humans, and then it is declared that it is precisely these properties that make humans a superior species, distinguishing them from all other species. Plato already pointed out the fallacious nature of this reasoning almost 2,400 years ago. According to Plato, a crane with the same intelligence and narcissism as humans would likewise divide living beings into two categories: cranes, objects of veneration, on the one hand, and all other living beings, reduced to the rank of "beasts," on the other.

But let us ignore this elementary error of logic for the moment and take a closer look at the list of human intellectual capacities, which does look impressive at first sight. I will not dwell on the human soul, a concept probably too vague to be tested by scientific facts. Ironically, however, note that the word "animal" comes from the Latin word *anima*, which means soul. Theologians and philosophers of the Christian era were so intent on stripping animals of any human-like attributes that they even sought to remove what defined them in the first place!

Let us start with self-awareness, which has long been claimed to be absent in animals. Although this concept covers a mental reality that everyone can easily perceive internally, it is much more difficult to define precisely what it is. Therefore, it is equally difficult to establish its presence or absence in other creatures in the absence of verbal communication with them. Self-awareness is a complex property, which is now known to have at least three dimensions: bodily self-awareness, social self-awareness, and introspective awareness (DeGrazia 2009).

It may seem surprising to talk about bodily self-consciousness when Christian religion and modern philosophy have so accustomed us to separating the mind from the body and glorifying the mind over the body. Yet it is the form of self-consciousness that plays the most important role in our lives because it shapes our identity in the face of the outside world. Bodily self-awareness is related to physical sensations; it allows us to perceive our body as distinct from the external world, as well as its internal state (hot, cold, hunger, pain, etc.). Many animals seem to have this primitive form of self-awareness, which makes perfect sense from an evolutionary point of view. Indeed, bodily self-awareness allows for a flexible and efficient response to multiple internal and external disturbances; it thus contributes to maintaining the bodily integrity of the organism and ensuring its survival.

Social self-awareness is the ability to conceive of ourselves as part of a social unit and to take account of differences in social status in our behavior. It is present at least in mammals with developed social behavior, particularly in primates and cetaceans, for whom it is an important asset considering the major threat that being ousted by a dominant conspecific in the event of inappropriate behavior represents.

Finally, introspective self-awareness is the awareness of one's own mental states, such as feelings, desires, and beliefs. Several recent experiments have demonstrated that primates possess awareness of their own mental states. For instance, in one such experiment, monkeys were taught to control a joystick to make choices on a computer screen. If they got the answer right, they were given food; if they got it wrong, they had to wait before they could play again, which they hated. They were then given the option of choosing an icon that allowed them to skip a test if they thought it was too difficult. They quickly learned to use this option wisely. This experiment therefore demonstrates that these primates were assessing their confidence in their ability to pass a test, thus demonstrating a form of introspective awareness (DeGrazia 2009). In another experiment, chimpanzees were asked to choose photographs of their faces representing their emotions when they watched videos of scenes evoking more or less positive or negative emotions. Without any prior learning, they correctly associated the photographs with their own emotional state, as measured independently by their body temperature (Parr 2001). This shows that chimpanzees are able not only to assess their internal emotional state, but also to choose an abstract representation that corresponds to it, demonstrating a well-developed introspective self-awareness.

Thus, there can no longer be any doubt that self-awareness is not unique to humans, but that it is present in many animals, at least in the elementary form of bodily self-consciousness, and sometimes even in the more elaborate form of introspective self-consciousness as we humans know it. But perhaps self-awareness is, after all, still too elementary a cognitive property to distinguish humans from animals. What about apparently more elaborate properties like empathy?

Empathy is often defined as the ability of a person to project onto another person his or her own mental state if he or she were in the situation experienced by the other person. In its most developed form, empathy implies a relatively high level of representation of the "self," as it requires the ability not only to mentally project the concept of "self" onto the other person, but also to anticipate how this "projected self" would feel in the situation experienced by the other person, and finally to assume that this mental state of the "projected self" equals the actual mental state of the other person (Schweitzer & Notarbartolo-di-Sciara 2009). Thus, empathy necessarily implies self-awareness: one cannot hope to understand the mental state of another person without first being able to assess one's own mental state. If it could therefore be demonstrated that an animal feels empathy toward another animal, this would also mean that this animal possesses self-awareness.

Numerous experiments with laboratory rats and mice have demonstrated the existence of empathic reactions toward fellow animals since the late 1950s (de Waal 2009). The first such experiment began with a classical setup in which rats were taught to obtain food by pressing a lever. The pressing of the lever by one rat was then associated with an electric shock sent to another rat, visible to the first. The rats quickly stopped pressing the lever to obtain food. This response was completely unexpected at the time. Why did these rats not continue to gorge themselves regardless of their companions who were writhing in pain near them? The interpretation given at the time was that the rats feared for their own well-being when they saw their fellow rats in distress. But this interpretation is obviously inconsistent: how could a rat that had never been subjected to any such experiment fear for its own welfare when it saw a fellow rat exposed to an unknown situation? It seems much more likely that a rat's distress would induce an emotional distress response in its fellow rats. Multiple experiments were then conducted to analyze the causes of similar responses in other animals, notably laboratory mice. The conclusion of these experiments is unequivocal: it is the pain or distress response of a known conspecific that causes the sensitization to pain or distress, regardless of how it is caused. Interestingly, this empathic response does not occur in the presence of an unfamiliar conspecific, indicating that it is not automatic.

Is this empathy? There is no doubt that the rats and mice in these experiments project an emotional state onto a fellow animal with whom they have made prior contact, which represents a form of empathy. But this does not necessarily mean that the rats, mice, primates, and other animals that have been shown to have empathic responses have a highly developed intellectual representation of the "self." As with self-awareness, it is now known that empathy is in the body before it is in the mind. Empathy probably has its roots in the synchronization of bodies: we are prompted to run, laugh, cry, or yawn when others do. Cognitive sciences are increasingly demonstrating that cognition itself is not based on purely intellectual processes, but that it involves the body and its sensations, in humans as in other animals. It is therefore quite natural that empathy has a strong bodily dimension. As ethologist and primatologist Franz de Waal (2009, 95) put it, "we unintentionally enter the bodies of those around us." Their movements and emotions resonate within us as if they were our own. Like self-awareness, empathy does exist in other animals, especially mammals. It is quite likely that it takes more developed forms in the human species. For example, imaginative empathy allows us to understand what the other person is feeling even when we cannot see him or her or when he or she is an imaginary character in a novel or film. Nevertheless, it is not imagination that mobilizes empathy. Empathy requires first and foremost an emotional commitment. Communication at the bodily level comes first; understanding follows.

Although this scientific knowledge undermines the hypothesis of a radical separation between humans and animals, the modern skeptic might say that, on balance, self-awareness and empathy are still too close to the body to do justice to the spiritual superiority of humans. Thinking, on the other hand, requires complex operations of the mind that should allow for a clear separation. What a disappointment it will be to our skeptic, then, to learn that even abstract thought is not unique to humans! Cognitive sciences have begun to take a serious look at this topic, and their results are a stark rebuttal to those old beliefs deeply rooted in our civilization: not only primates, but even bees possess the building blocks of abstract thought. Of course, defining thought is again a delicate operation, but scientists recognize that it includes at least the following components: on the one hand, distinct states of belief and desire that interact with one another and with perception to guide behavior; on the other hand, a structuring of belief states into elementary components—the concepts—that can be recombined in various ways (Carruthers 2009).

Ingenious experiments have recently shown that bees do have belief and desire states that interact with one another to guide their flight behavior, and that their belief states involve distinct symbols that refer to substances, spatial cues, distances, and directions. Moreover, these symbols are real concepts, as bees can combine them in many different ways to elaborate flight-related thoughts. For example, the concepts of "nectar," "pollen," "distance," "direction," and "hive" can be combined indifferently to produce thoughts such as "the hive is 200 meters north of the nectar," "the nectar is 200 meters west of the hive," or "the pollen is 400 meters north of the hive" (Carruthers 2009). And we know that the bee does use these thoughts to guide its flight.

In a famous passage in *Capital*, Karl Marx claimed that "what distinguishes the worst architect from the best of bees is this, that the architect raises structure in imagination before he erects it in reality" (Marx 1965 [1867], 728). It seems, then, that the bee is much less stupid and different from humans than Marx and his contemporaries imagined, victims as they were of the anthropocentric mirage characteristic of modern civilization. There is little doubt that overall a human has more diverse intellectual capacities and more complex thinking than a bee, but this is only a difference in degree, not in kind. In fact, it would not be surprising at all if bees had higher intellectual capacities than humans for specific tasks such as distance assessment and spatial orientation, as these play a more crucial role in their daily lives than in the daily lives of humans. A recent study showed that bees can count and even have the concept of zero (Howard et al. 2018), an abstract concept that had long been regarded as one of the greatest feats of the human mind. We are now discovering that even plants—which have traditionally been considered the most "stupid" organisms because they have no brains—have remarkable intelligence and communication skills, albeit in very different forms from our own (Mancuso 2018).

Another intellectual ability that has long been considered the prerogative of humans and their abstract thinking is known as "theory of mind." Theory of mind refers to the ability of an individual to attribute mental states, such as intentions, goals, beliefs, and knowledge, to other individuals, in other words to know what they know, intend, or believe. Until recently, it was thought that this ability existed only in humans and that it developed in children around the age of four. This belief has, once again, been overturned by recent scientific studies using ingenious eye movement–tracking technology in place of the traditional verbal response choice experiments, which are inappropriate for young children and animals. These studies have shown that primates, like young children, are able to anticipate the behavior of a person searching for a hidden object where they themselves know it is not. Our primate cousins are therefore able, like us, to know that other individuals hold false beliefs (Krupenye et al. 2016).

If self-awareness, empathy, and thought are present, at least in primitive form, in animals other than humans, it may be less surprising to learn that the same is true of language and culture, which for a long time were also considered to be the prerogative of the human species. Admittedly, scholarly debates continue to rage over these issues because of their strong emotional charge, as they do over self-awareness, empathy, and thought. But, if you look at it, these debates boil down to definitional problems that are, after all, quite secondary. Indeed, it is always possible to arrange to choose a definition that applies only to the human species—this is the crane syndrome highlighted by Plato. If we free ourselves from the obsession with finding a characteristic that separates humans from the rest of the animal kingdom at all costs, we can only be intrigued and amazed by the unsuspected skills of animals, which share much more with us than we previously thought.

Even in humans, language is commonly defined as the ability to express thought and communicate through a system of signs (whether vocal, gestural, graphic, tactile, olfactory, or other). If we stick to this common definition, there is again no doubt that many other animals have language. Take the case of bees. We have seen that bees have a form of thinking that involves the concepts of distance, direction, and quality of a food source. We also know from the classic work of Karl von Frisch that bees communicate information about the distance, direction, and quality of a food source to each other through a highly symbolic "waggle dance." The combination of these two skills is precisely what is commonly defined as language.

Examples of language skills abound in primates and some birds. It is well known that chimpanzees can be taught human sign language. Some experiments have even shown that chimpanzees understand the abstract idea of category; they can use the sign "dog" to refer to any kind of dog or "shoe" to refer to different kinds of shoes. Better still, some chimpanzees spontaneously create unlearned combinations of signs to express new ideas. Further away from humans, parrots also have remarkable linguistic skills. Contrary to popular belief, parrots do not just stupidly repeat what they hear; they think, use abstract concepts, and can communicate their thoughts using their vocal apparatus, which allows them to establish verbal communication with humans. Perhaps the best-known example is Alex, the African gray parrot trained and studied by ethologist Irene Pepperberg as part of a scientific project on parrots' ability to understand human language (Pepperberg 2009). Alex could identify and name about fifty different objects, seven colors, and five shapes; he could count to six and understand concepts such as "bigger than," "smaller than," "same as," and "different from." Most importantly, he seemed to understand perfectly the meaning of what he heard and said. He answered any questions about shapes, colors, materials, and numbers correctly, which means that he understood not only the meaning of the words for a particular color or shape, but also the concept of color or shape itself. He also knew what he wanted and communicated it to the experimenter. For example, when he was tired of the experiments, he would say, "Wanna go back" (to his cage), and if the researcher displayed irritation, he would try to defuse it by saying "I'm sorry." Although these examples of chimpanzees and parrots being trained to express their thoughts using human language are somewhat artificial, they do demonstrate that these animals have cognitive and linguistic abilities that are very similar to those of humans, and therefore that these abilities are not unique to humans. These examples of animals "speaking human" even contain a good dose of irony, because, to my knowledge, no human has yet managed to speak chimpanzee or parrot.

What about culture? Until recently, culture was thought to be the most significant difference between humans and other animals, so much so that culture and nature are still commonly contrasted, as if they were two separate worlds. Only humans, it seemed, possessed the ability to shape their behavior on the basis of a set of shared knowledge and practices transmitted within a social group, whereas animals had an innate behavior, fixed once and for all by their instinct and genes. We now know that this is not the case. On the one hand, humans are much more similar to each other, beyond their cultural differences, than appears at first sight—we will come back to this issue in Part II of this book. On the other hand, animals also shape their behavior according to their social environment, thus creating cultural differences between populations that are passed on from generation to generation.

The first examples of cultural learning and transmission were observed in Japanese macaques by primatologist Kinji Imanishi and his colleagues in the late 1940s. These researchers noted significant differences in social norms and feeding behavior between different groups of macaques, which they attributed to social learning. They then witnessed firsthand the social learning of a new feeding behavior when a young female washed sweet potatoes before eating them, a behavior that had never been observed before. This behavior was quickly adopted by the young female's playmates, then by her mother, and finally spread to the entire colony. Since then, examples of culturally learned and transmitted behavior in social animals have multiplied, particularly in primates and cetaceans. For example, chimpanzee populations in Africa differ in dozens of learned behaviors, including the use of leaves, branches, and stones for communication, play, or foraging. The use of these tools is specific to each population and transmitted within it through a mixture of imitation and social learning. Similarly, whales and dolphins are organized into groups with their own vocal dialects. A recent study even succeeded in experimentally initiating a new feeding cultural trait in wild vervet monkeys (van de Waal et al. 2013).

It is therefore becoming increasingly clear that, far from being in opposition to it, culture belongs to nature. The same applies to morality. The modern conception of morality is strongly rooted in rationalism, in particular in Kant's philosophy, which bases moral behavior on a conscious choice made by rational human beings. But this view has been challenged by recent studies in neurosciences, human psychology, and animal psychology (Hauser 2006). This work has showed that moral decision-making is primarily driven by emotions; it activates parts of our brain that go back to the transition from cold-blooded reptiles to the caring, loving, infant-feeding mammals that we are (de Waal 2005). There are several famous medical cases of people who had suffered deep damage to the frontal lobes of the brain after a serious accident. Perhaps surprisingly, these people had retained all their intellectual faculties, but they had a strongly disturbed emotional behavior and were unable to make decisions, especially of a moral nature. The study of these medical cases revealed that the damaged regions of the frontal lobes were areas of the brain where reasoning and emotional perception and expression processes are integrated (Damasio 1994).

Thus, paradoxically, emotions underlie morality. Rationalizations often come after the fact, when we have already reacted according to our species' predispositions. This observation may seem surprising because we have been educated to believe otherwise, but many moral choice experiments support it. Hypothetical examples can illustrate quite simply the limits of logical reasoning in our moral choices. Imagine, for example, an enlightened dictatorial political leader who decided to kill anyone who might carry the coronavirus responsible for the recent Covid-19 pandemic. By doing so early enough, he would probably have killed far fewer people than the pandemic did, thereby saving many lives. Such a policy would be perfectly rational, and yet it would be met with disgust and would be considered immoral by most people. This example shows that, contrary to the doctrines of rationalism and utilitarianism, humans do not work for the well-being and happiness of the many if their behavior violates the fundamental inhibitions of our species. Note that the power of emotions in moral behavior does not mean that rationality is unimportant. Rational thinking allows organizing our emotional responses in a coherent and systematic way, thereby cementing human communities around shared rules of action and behavior. These are very important skills for highly social animals like humans.

If morality is fundamentally rooted in emotions rather than in reason, it seems difficult to exclude rudimentary forms of morality in animals. The more we study animals and humans, the more we are struck by the similarity of their behaviors and skills. A recent study, for example, revealed the existence of post-traumatic stress disorder in elephants. Young elephants that have witnessed the killing of their parents show severe behavioral disorders in adolescence: they become abnormally aggressive and sometimes proceed to wantonly slaughter other animals. Hyperaggression, however, disappears in the presence of adult males (Bradshaw et al. 2005). The similarity with humans is striking: it is as if young elephants, traumatized by the killing they witnessed, lost their moral bearings in the absence of strong family or social ties. It is possible that the role played by intellectual judgment is more limited in elephants than in humans in the choice of their behavior—although in reality we just do not know. In any case, the result is the same: both humans and elephants are capable of being violent, but a secure emotional climate, combined with a probably varying degree of reasoning about the consequences of their actions, leads them to curb this violence and find a more peaceful form of expressing their needs.

A final feature that has long been considered the prerogative of human intellectual genius is the making and use of tools. We now know that many other animals, especially primates and birds, use tools to search for food. Since Jane Goodall's pioneering work on the behavior of chimpanzees in the wild, we have known that they use a wide range of tools. The making and handling of these tools sometimes requires a complex chain of operations that demand a high degree of anticipation, coordination, and manipulation (Beyries & Joulian 1990). Termite fishing provides a good example. Chimpanzees first inspect termite mounds and locate their entrances before the rainy season to prepare for fishing. At the time of termite swarming, which takes place at the beginning of the rainy season, they select and shape thin wooden sticks, carry them to the previously identified termite mounds and then use them for the actual fishing. This consists of carefully inserting a stick into a hole in the termite mound, waiting for the termites to cling to it by biting, then removing the stick from the hole and eating the clinging termites. In the intentionality and high complexity of the tasks involved, termite fishing is not fundamentally different from human tool use. Other examples of the same type can be found in birds. For example, some Darwin's finches in the Galapagos Islands use cactus spines as skewers to prick insects from tree branches and eat them. Not content with just using the thorns they find, they modify their size and shape to make them more effective tools.

Ants may be even more impressive, not so much because of the complexity of their individual behavior, but because of the social organization of their livelihood production techniques. Ants collectively raise and exploit other living things for their livelihood—in other words, they invented agriculture long before humans did. Many species of ants raise aphids in the same way as we raise cows or sheep: they herd them, protect them from predators and parasites, and milk them for a sweet substance called honeydew. Mushroom ants in the tropics, on the other hand, cultivate a species of mushroom they use as food. The fungus is carefully cultivated in purpose-built gardens; the ants supply it with cut leaves for growth and even carry a filamentous bacterium that produces antibiotics to protect the cultivated fungus from other parasitic fungi (Currie et al. 1999).

Wherever we look, we see that science is systematically destroying, one by one, all the old prejudices that claimed to make humans a species apart, at the top of creation. Only a few nostalgic philosophers still insist on defending the idea that there is a radical divide between humans and animals (Ferry 1992; Bimbenet 2011). As I mentioned above, it is always possible to find a difference between humans and any other animal so great that this difference seems infinite and therefore qualitative; but it is also possible to find a difference as great, if not greater, between two animals belonging to different species.

The relativization of the place of humans in the world and in the history of life brought about by modern biology may be experienced by some as a painful loss of illusion, just as was the replacement of the geocentric conception by the heliocentric conception of the universe a few centuries ago. But above all, it invites us to a magnificent opening to the world around us. Essayist Jeremy Rifkin (2009, 104) speaks of this opening in beautiful words:

> What scientists are finding is that human beings share a much richer history with our fellow mammals than previously thought. We now know that mammals feel, play, teach their young, and show affection and, at least some species, have a rudimentary culture and express primitive empathic distress. We are finding kindred spirits among our fellow creatures. Suddenly, our sense of existential aloneness in the universe is not so extreme. We have been sending out radio communications to the far reaches of the cosmos in the hopes of finding some form of intelligent and caring life, only to discover that what we were desperately seeking already exists and lives among us here on Earth.

Although *Homo sapiens* is not a creature apart, although it is part of the tree of life and has much more in common with other living beings than has been acknowledged so far, it is nevertheless a species distinct from others and therefore unique in its own way. Indeed, like any species, humans have a unique combination of characteristics that define them as a species. Now, it is essential to understand the uniqueness of the human species to understand the conflicting relationship it has established with the rest of nature.

A set of biological characteristics distinguishes humans from other primates, such as standing, a prehensile hand, the development of the cranium, and a particularly long period of care for the young. There is no doubt that the combination of these interrelated characteristics played a key role in the hominization of early humans. But what has made *Homo sapiens* such a special animal that it comes to conceive of itself as alien to the world of which it is a part is apparently the development of the prefrontal cortex of the brain, which is the seat of the so-called higher cognitive functions such as reasoning and language. As I pointed out earlier, thinking and language as such do not distinguish humans from other animals. But the development of the prefrontal cortex has increased the ability of humans to create associations between concepts, leading to the acquisition of a new capacity for creative supposition, that is, to entertain thoughts that explore the possible without necessarily aiming at truth or adequacy with experience (Carruthers 2009). Some believe this new capacity for creative supposition stems from the importance of tracking large animals, when early humans developed the activity of hunting and moved from a frugivorous to an omnivorous diet. Indeed, tracking requires the ability to read and interpret animal tracks as visible signs of an invisible reality (Morizot 2018). In any case, the human capacity for creative supposition does not seem to have any equivalent in the rest of the animal kingdom. In other words, we seem to be the only animal species to tell stories about things that do not exist—and, in the process, to tell ourselves stories, to mystify ourselves.

The significance of this biological innovation cannot be underestimated. Although it represents only a minor change in our biological makeup, the ability to make fiction has opened up new and unsuspected horizons for the human species. Narratives and fictions allow us not only to imagine things that do not (or do not yet) exist, but also and above all to do so collectively. Shared narratives and fictions are a powerful lever to bind together human communities around common goals and actions. They give humans an unprecedented ability to cooperate in large numbers and thereby increase their collective power. As historian Yuval Harari (2011, 42) rightly notes: "One on one, even ten on ten, we are embarrassingly similar to chimpanzees. Significant differences begin to appear only when we cross the threshold of 150 individuals, and when we reach 1,000–2,000 individuals, the differences are astounding. (. . .) The real difference between us and chimpanzees is the mythical glue that binds together large numbers of individuals, families and groups. This glue has made us the masters of creation."

Shared narratives and the development of sophisticated languages that accompanied them greatly contributed to increasing the internal cohesion of human groups, as well as the divergence and conflicts between groups, thus paving the way for an extraordinary cultural diversification. This accelerated cultural evolution in turn reinforced the importance of shared narratives in group identity. Every human society is characterized first and foremost by its own vision of the world, by a narrative of its origins, by a set of collective fictions that governs how its members interact with one another and with the other animate and inanimate beings that constitute their environment. These collective fictions define the contours of its mode of action in the world, its social organization, and the technical progress that can take place in it (Descola 2005). They give rise to myths, religions, utopias, science, and thus, directly or indirectly, to everything that makes up modern humans.

But this biological innovation does not only have advantages. Fiction and culture have come to assume such an importance in the development of human beings that they tend to overshadow their fundamental nature and needs. *Homo sapiens* seems to be the only species that has difficulty in seeing itself as a species (Maslow 2006 [1971]). Cats seem to have no trouble being cats; they show no signs of wanting to be dogs instead—their instincts seem to be perfectly clear. By contrast, our biological essence and instincts are harder to perceive. Extrinsic learning often relegates our deepest impulses to the background, at least in our mind. We have great difficulty in accepting ourselves as we really are; we often dream of being something we are not, and we may even spend our entire life trying to make this dream come true. The stories we tell ourselves, individually and collectively, often take precedence over our most basic needs, to the point that we ignore or deny our needs deliberately, resulting in inner conflicts, neuroses, and dreams of domination. The separation of humans from nature is first and foremost a separation of humans from their own nature. Stories then turn into History, for better or for worse.

2

A brief history of the divorce between humans and nature

Humans are, in many ways, animals like any other. Why, then, have they come to see themselves as alien to nature, that is, to the rest of the world around them? Why do modern humans display such a sense of singularity, superiority, and even rejection toward nature?

Any question about the causes of a phenomenon invites an infinite number of answers. Every phenomenon is the result of an extraordinarily complex set of processes, more or less close or distant in space and time, that contribute to its manifestation here and now. Young children understand this intuitively. Every parent has experienced being overwhelmed by a flood of "Why?" from their child when they reach the age of three, only to find, bewildered, that no answer satisfies their curiosity. Each answer is invariably followed by another "Why?" and so on until the parent is tired of answering. This can be interpreted as a game, and it is, for the child often derives some pleasure from it, but it is a serious and salutary game because it invites us to rediscover with the child inside us that causality is infinite and that our automatic answers are nothing more than convenient ways of simplifying our daily lives.

So, I do not believe that there is a single cause for the divorce between humans and nature that characterizes modern society. Many books have examined this question and have proposed a wide range of more or less compelling hypotheses. It is not my aim here to examine these hypotheses in detail. In fact, many of them are complementary and difficult to distinguish. What interests me is rather to understand the cluster of factors that contributed to the emergence of the modern relationship between humans and nature and that contributes to its perpetuation today. This cluster exists because the separation between humans and nature has been established gradually, over the course of a long if tumultuous history.

I need to start with the earliest history, namely the evolutionary history of the human species. There is a fairly widespread idea, especially among my biologist colleagues, that the destruction of nature by humans is the expression of an innate destructive behavior of the human species. If this idea were

Nature That Makes Us Human. Michel Loreau, Oxford University Press. © Oxford University Press 2023.
DOI: 10.1093/oso/9780197628430.003.0003

correct, it would provide a simple answer to the question posed and save me the trouble of writing this book. Jared Diamond, who has written remarkable books on the development and decline of human societies in relation to their natural environment, is one of the biologists who have popularized this thesis. In his book *The Third Chimpanzee*, Diamond (1992) offers a fascinating evolutionary perspective on the human species, but also a rather pessimistic view of human nature since, according to him, humans have always engaged in genocide and the destruction of their environment and biodiversity. More recently, psychologist Thierry Ripoll (2022) has even gone so far as to argue—against a great deal of scientific evidence to the contrary—that it is the biological structure and functioning of their brains that drive humans to desire more and more. Thus, according to him, modern capitalism and the contemporary destruction of nature would be inevitable consequences of the evolution of the human brain.

There is no longer any doubt that humans have had a destructive impact on their natural environment for quite a long time (Johnson et al. 2017). Perhaps one of the most dramatic examples of this destructive impact is the extinction of several thousand endemic bird species on the Pacific islands following human colonization over the past 30,000 years (Steadman 1995). Does this mean that humans are destructive by nature? It is interesting to note that, on the basis of detailed observation of contemporary hunter-gatherer peoples, some ethnologists and ecologists have come to the diametrically opposed conclusion that most hunter-gatherers are, on the contrary, particularly respectful of their natural environment and the living beings that inhabit it. So, who is right?

In truth, humans are neither intrinsically destructive nor intrinsically respectful of nature. Like any living being, humans have needs to satisfy, and they adopt behaviors or strategies that are more or less adapted to the satisfaction of these needs. All life is a continuous process of creative destruction and destructive creation. Every living being uses and modifies physical, chemical, and biological elements of its environment to grow and sustain itself; in doing so, it destroys the form in which these elements were present to create and maintain its own form and function. Thus, the plant uses solar energy and a number of nutrients in the soil and in the atmosphere to create and maintain its own plant form and function; the herbivore consumes the plant to create and maintain its own herbivore form and function; the carnivore eats the herbivore to create and maintain its own carnivore form and function; and the bacterium uses the dead bodies and residues of the plant, herbivore, and carnivore to create and maintain its own bacterial form and function. Each of these living beings at once creates itself, destroys the elements of its environment that serve to create it, and transmits these elements in new form to the next link in the chain of life. It is therefore futile to try to separate in the abstract what is creation, what is destruction, and what is respect for life. It is all a matter of context, of balance or imbalance, of direct or indirect consequences that are more or less beneficial or harmful to the various links in the chain of life.

That humans are not fundamentally, congenitally destructive will be obvious to anyone who has raised a child in an atmosphere of caring and self-fulfillment. Children are spontaneously curious and attracted to other living beings, especially animals, with whom they often establish a strong empathic connection. I have not known any child who spontaneously enjoyed destroying life; the few children I have known who engaged in destructive behavior did so as a tragic expression of a lack of love or recognition they were suffering. If their fundamental needs are fully met, there is no reason why human beings should not transform their spontaneous empathy for other living beings into a respectful attitude toward nature. If their fundamental needs are not met, however, they will develop strategies to try to satisfy at least their most basic needs, even if it means destroying their environment and thus the very conditions for their long-term survival. Thus, the real question is this: what are the ecological and cultural contexts that explain, on the one hand, the destructive impact of humans during the colonization of the Pacific islands and, on the other hand, the respectful attitude toward the rest of nature of many contemporary hunter-gatherer peoples?

Recent studies suggest that the destructive impact of human colonization of the Pacific islands was, in fact, much more limited during the Paleolithic than previously believed. Many of the documented endemic bird extinctions on the Pacific islands probably occurred in the Neolithic period as a result of deforestation and the development of agriculture (Steadman 1995). A recent comprehensive analysis of archaeological and paleontological records concluded that there was no evidence for widespread species extinctions on islands following human colonization during the Paleolithic globally (Louys et al. 2021). It is not until the Neolithic, when large-scale changes in human social organization, technology, dispersal, and demography took place, that humans visibly affected island ecosystems. We will return later to the transition from the Paleolithic to the Neolithic as a key historical event that generated major changes in the relationship between humans and nature. For the time being, let us simply observe that the endemic bird extinctions that occurred on the Pacific islands following human colonization do not support the hypothesis that humans have had a systematically destructive impact on their environment since the beginning of their existence as a species, and thus that they are congenitally destructive. Many other examples of species extinctions during the Paleolithic that have been attributed to humans also show no evidence of a causal relationship with destructive human behavior. In particular, there is a long-standing scientific debate about the causes of the extinction of large mammals and other slow-growing animals between fifty and ten thousand years ago on all continents except Africa. A popular hypothesis is that humans drove large animals to extinction through hunting after colonizing new regions or continents. This hypothesis, however, seems at best an oversimplification. Instead, recent studies suggest that large mammal extinctions in the late Quaternary resulted from a combination of factors, including climate change and a wide range of human direct and indirect impacts (Koch & Barnosky 2006).

Paradoxically, the example of the Pacific islands could even be used in support of the opposite hypothesis that Paleolithic hunter-gatherers were relatively non-destructive. Indeed, the colonization of new territories by humans constitutes, in itself, a particularly favorable context for adverse consequences for endemic fauna because it brings together species that were not previously in contact and thus have not had the opportunity to adapt to each other during their evolutionary history. The Pacific islands provide a particularly good potential example of this rule. First, these islands are small and highly isolated, two factors that are known to greatly exacerbate the impact of exotic predators, whether human or non-human. The extinction of endemic species as a result of the accidental or deliberate introduction of a predator or a disease to remote islands is a well-established phenomenon in ecology. Second, most of the endemic species that became extinct there were rails that had lost the ability to fly in the absence of predators prior to human colonization. As a result, they were particularly vulnerable and defenseless in the face of any sort of predation. It is not difficult to imagine that even moderate hunting pressure could lead to the rapid disappearance of their populations under these circumstances, which is probably why the hypothesis that humans caused bird extinctions during the Paleolithic was accepted uncritically. Given that human colonization was so highly conducive to the extinction of flightless endemic rail species on these remote Pacific islands, what looks surprising, in fact, is not so much that there were extinctions, but rather that there were apparently so few extinctions before the Neolithic.

It is quite possible that Paleolithic hunter-gatherers developed an attitude of respect toward nature as observed in a number of contemporary hunter-gatherer peoples, despite, or perhaps as a result of, the adverse impacts they may have had on their natural environment. This hypothesis would make sense from an evolutionary point of view. It may be useful to recall here that the genus *Homo* appeared nearly 3 million years ago and that our species *Homo sapiens* appeared nearly 300,000 years ago. Since the Neolithic did not begin until about 11,000 years ago and did not spread until much later, this means that most of the evolutionary history of the human genus and of the modern human species took place in the Paleolithic. Therefore, most of the traits that define human nature are inherited from a time when only the hunter-gatherer lifestyle existed. Furthermore, it seems that for most of the Paleolithic period, the human lifestyle was essentially that of a gatherer, with hunting only appearing at a later stage. It is quite possible that the spread of hunting led some human populations to temporarily overexploit the large species of mammals, birds, and reptiles that offered them particularly rich and accessible resources, leading some of them to extinction, especially when colonizing new territories. The colonization of new territories, however, is a relatively rapid event in the evolutionary history of hominids—on the order of a few hundred or thousand years, out of an evolutionary history of about 3 million years. For the rest of their evolutionary history, the human genus and species lived in a relatively stable environment (at least on the scale of a human lifetime, not counting longer-term climatic and other environmental changes), upon which humans were closely dependent for subsistence and survival. In this context, an in-depth knowledge of the living beings around them was essential. Since humans did not have access to sophisticated instruments and technology at that time, in-depth knowledge was based on patient and detailed observation of plants and animals. An attitude of acceptance and respect is an obvious asset under these conditions.

Anthropological studies of contemporary hunter-gatherer peoples have profoundly changed our view of human history and prehistory over the past decades. These studies have shown in particular that hunter-gatherers, who have preserved a "primitive" way of life close to that which probably prevailed during the Paleolithic period, enjoy a remarkable quality of life in many respects. This high quality of life contrasts strikingly with the modern cliché of the "caveman," who is supposed to live like a brute and be constantly threatened by famine and disease. Anthropologist Marshall Sahlins (2017 [1972], 14, 36) argued that the exact opposite is true. According to him, the "subsistence" economy of primitive societies is, in fact, a society of affluence, whereas modern society is a society of scarcity: "A good case can be made that hunters and gatherers work less than we do; and, rather than a continuous travail, the food quest is intermittent, leisure abundant, and there is a greater amount of sleep in the daytime per capita per year than in any other condition of society." "The world's most primitive peoples have few possessions, *but they are not poor*. Poverty is not a certain small amount of goods, nor is it just a relation between means and ends; above all it is a relation between people. Poverty is a social status. As such it is the invention of civilization. It has grown with civilization, at once as an invidious distinction between classes and more importantly as a tributary relation—that can render agrarian peasants more susceptible to natural catastrophes than any winter camp of Alaskan Eskimo."

What modern civilization has reified in the form of a "nature" external to humans is, in the eyes of the hunter-gatherer, a world populated by creatures animated by the same life breath as he or she is and to which he or she belongs inseparably (Descola 2005). There is no hierarchy of living beings with humans at the top. Most animals are conceived as persons with souls, which gives them attributes identical to those of humans, such as self-awareness, intentionality, emotional life, and respect for ethical precepts. For Native Americans (of both North and South America), hunting is conceived as a social interaction with beings who are fully aware of the conventions that govern it. It is therefore by showing respect to hunted animals that one ensures their complicity, without which hunting would not be possible. Furthermore, the nomadic lifestyle of many hunter-gatherer peoples leads them to exploit a limited amount of resources in any one place before migrating to other more suitable areas (Sahlins 2017 [1972]). Thus, neither their worldview nor their way of life predisposes these peoples to exploit other living beings in a brutal and excessive manner—although, of course, they offer no guarantee that any form of overexploitation be excluded, especially when colonizing new territories.

This is not to indulge in the old myth of the Golden Age. The hunter-gatherer way of life has aspects that may seem cruel to modern humans—at least to those who enjoy all the benefits of modern comfort. In particular, births, deaths, and thus population size are largely regulated by the constraints of nomadism. Individuals who are either too old or too ill to participate in long-distance walking to new territories are unlikely to survive. The worldviews and beliefs of hunter-gatherers, which are extraordinarily diverse, are, like those of modern humans, stories that they tell themselves and that reflect particular historical trajectories. Thus, they should not be seen as truths against the excesses of modernity. But in many respects, it does seem that the hunter-gatherer way of life that prevailed in the Paleolithic was an age of relative affluence and simplicity, of which humanity may have retained a nostalgic memory in the idealized form of the myth of the Golden Age. This may explain why some populations decided to convert back to gathering and hunting when conditions permitted. For example, a recent genetic study revealed that the last hunter-gatherers of Madagascar, the Mikeas, actually originated from a population of farmers and herders who had converted to a hunter-gatherer lifestyle (Pierron et al. 2014). Other hunter-gatherers, such as the Hadza of Tanzania, have long refused to adopt any agricultural practices, claiming that this would entail too much work, whereas gathering and hunting effortlessly provided everything they needed (Sahlins 2017 [1972]).

If the hunter-gatherer way of life that has prevailed during most of human history is largely incompatible with a separation between humans and nature, where, then, did this separation come from? There is some disagreement as to the ultimate origin of this separation, but much agreement as to its subsequent historical development. Ecofeminism places the origin of human domination over nature in patriarchy (Mies & Shiva 2014), as the domination of men over women leads directly to the domination of humans over nature insofar as women symbolize the "natural" side of humankind, particularly through their role in procreation. Others place it in the emergence of social hierarchies, which probably preceded patriarchy, with the domination of humans over nature being an extension of the domination of humans over other humans (Bookchin 2010). Still others emphasize factors such as animal domestication, as humans subjugated large animals and transformed them from objects of respect and admiration into mere possessions (Mason 2005), or the emergence of alphabetic writing, which gradually disconnected humans from their direct sensorial experience of the world (Abram 1996). In any case, all of these factors were inextricably intertwined in the Neolithic revolution that began in the Middle East about 11,000 years ago and that has largely shaped modern Western civilization, one of whose unique features is precisely the separation of humans from nature (Descola 2005).

The Neolithic revolution was particularly profound and rapid in the Middle East, where it was accompanied by the domestication of a large number of plant and animal species for the production of food resources, the emergence of agriculture and pastoralism, and the transition from the nomadic lifestyle typical of hunter-gatherers to a sedentary lifestyle. But this upheaval in the relationship between human populations and their environment set in motion a wider revolution in the whole of social life. In particular, it led to considerable increases in birth rate and population density, which were then offset by an equally significant increase in mortality following the appearance of epidemic diseases and wars, for which a high population density is a fertile breeding ground. Finally, it favored the emergence of a complex social organization, hierarchically divided into social classes and regulated by an increasingly powerful state.

The modern ideology of progress presents human history as a steady progression toward a better life free from natural constraints, but all the evidence suggests that this has not been the case. On the contrary, the Neolithic revolution seems to have been an extraordinarily painful bifurcation in human history. Certainly, a small minority of the wealthy and powerful benefited from it—history as it is traditionally taught is essentially the history of this privileged minority. But for the majority of the men and women who wrote history with their sweat and blood, the adoption of agriculture was accompanied by an appalling deterioration in their quality of life. Forced labor, mobilization in war expeditions, epidemic diseases, and malnutrition were their fate. These scourges created a situation that would be regarded today as a complete disaster in terms of public health and quality of life, and probably caused deep and lasting traumas in collective human consciousness.

Paleopathology, a recent discipline that studies the signs of disease in human fossil remains, provides objective data that demonstrate this dramatically. For example, the average height of hunter-gatherers living in Greece and Turkey toward the end of the Ice Age was 1.78 m for men and 1.68 m for women. After the adoption of agriculture, around 4000 BC, average height had dropped to 1.60 m for men and 1.55 m for women, a decrease of 18 cm for men and 13 cm for women! Although average height slowly increased again thereafter, and then more rapidly in the past century with general improvements in diet and health, today's Greeks and Turks still have not regained the height of their hunter-gatherer ancestors who lived in the region (Diamond 1992).

Another example of the devastation caused by the adoption of agriculture is provided by the study of thousands of Native American skeletons exhumed from burial mounds in the Ohio and Illinois river valleys. Maize, grown in Central America for thousands of years, became the basis of intensive agriculture in these valleys around the year 1000. Until then, the skeletons of the hunter-gatherers show excellent health. After this date, they reveal a wide range of pathologies linked to malnutrition. The average number of dental caries per adult increases from less than one to nearly seven; tooth loss and dental abscesses become common; defects in the milk teeth of young children reveal acute malnutrition in pregnant and lactating mothers; the frequency of anemia quadruples; tuberculosis becomes an epidemic disease; half of the population suffers from syphilis or yaws; two-thirds of the population develop osteoarthritis or degenerative diseases; and mortality rises sharply at all ages.

There is little doubt today that agriculture, which enabled humanity to make a leap forward in terms of population size, was simultaneously a giant leap backward in terms of quality of life. Forced to choose between limiting human population growth and increasing food production, humankind opted for the second alternative, which paradoxically resulted in famine, war, social division, and tyranny (Diamond 1992). The Neolithic revolution, however, was not the result of a deliberate choice, but rather of social forces that imposed themselves gradually, leaving little room for free choice. Once the transition to a sedentary agricultural way of life was underway, agricultural societies had a considerable selective advantage over hunter-gatherer peoples living in the same region because of their high population density and the division of labor and war power that agricultural surpluses allowed them to develop. Agricultural societies quickly became too powerful for hunter-gatherers to stand a chance against them.

This painful transition profoundly altered people's vision of their relationship with the rest of the world. The nomadic hunter-gatherer lives and perceives the living world around him as being populated by beings who take care of his needs. He knows neither production as such, nor surpluses, nor cumbersome possessions that would hinder him in his movements. In contrast, the sedentary farmer is bound hand and foot to his land and to the work of his land to ensure his subsistence. The very nature of his work tends to place him in opposition to the rest of nature. He has to clear his land to cultivate it and then make constant efforts to get rid of the wild plants and animals that threaten to ruin his crops and return his land to its original state. He also has to control and confine the animals he has domesticated to prevent them from escaping and returning to the wild or being eaten by predators. Indeed, the entire agricultural system is based on the subjugation of nature and the domination and exploitation of other living beings (Serpell 1986). Far from contributing to greater security, this subjugation of nature was also accompanied by increased insecurity of human existence. For the first time in the evolution of the human species, farmers had to deal with alternating periods of abundance when they had too much to eat and accumulated surpluses, and periods of dearth when they had too little to eat. It is estimated that early farmers had a good harvest only about every seven years (Mason 2005). It is therefore not surprising that agricultural societies gradually developed a culture based on fear, anger, and resentment toward a hostile and capricious nature that only allowed them to eat their fill occasionally. The paradox is that this hostile nature is the one they themselves created by adopting a sedentary agricultural lifestyle and thus becoming more dependent on the vagaries of local environmental conditions. In contrast, by moving around as resources become available, hunter-gatherers experience a largely benevolent nature that provides for them without much effort.

The Neolithic revolution in the Middle East was characterized not only by the emergence of agriculture, but also by the domestication of large animals. The dog seems to be the only animal species to have been domesticated in the Paleolithic; all the others were domesticated in the Neolithic or later, and particularly in the Middle East. The domestication of large mammals such as cattle and horses could not have taken place without the use of force and the establishment of a relationship of submission between animals and humans. This submission is not only physical, it is also psychological: the animal must lose its soul ("animal" comes from *anima*, soul in Latin), that is, its personality, its independent activity and will, in order to accept to submit to humans. The reduction of large animals from conscious animate beings respected by hunter-gatherers to mere objects of property by pastoralists probably played a significant part in consolidating the view that humans are outside and above nature in Middle Eastern civilizations, the ancestors of modern Western civilization (Mason 2005). Interestingly, it also contributed to the emergence of capital and market wealth. Indeed, livestock was one of the earliest forms of market wealth—the term "capital" is derived from *capita*, meaning heads (of cattle) in Latin. The reduction of large animals to inanimate objects of property was a decisive step in the emergence of capital as well as in the emergence of human domination over nature. This is of course no coincidence for, as we shall see later, capital is the most complete expression of human domination over nature.

Finally, the Neolithic transition was also accompanied by the generalization of the division of labor and the stratification of human societies into social classes. The appropriation of agricultural surpluses allowed the emergence of privileged social classes freed from the labor of food production, centralized states, and armies geared to conquest. The division of society into classes necessarily implies relations of domination between classes. Thus, entire societies have been created which are impregnated, in their practice as well as in their ideology, by the domination of humans by other humans. This schism within human societies could only contribute to reinforcing the idea that humans and nature are also in a relationship of domination. Indeed, if, in the eyes of the dominant class, the human species is "naturally" divided into masters and slaves, and the manual labor involved in exploiting nature falls to the latter, nature must logically be even lower than the slave on the scale of power relations between living actors. The crystallization of this worldview into a coherent ideology was therefore, in a sense, only a matter of time.

While the conception of nature as external to humans, to be dominated and exploited, has its roots in the Neolithic transition in the Middle East, particularly with the emergence of agriculture, animal domestication, and social classes, it was only much later that it was fully developed and spread to other parts of the world. Many agricultural societies have long continued to view the cosmos as a great interconnected whole, of which humans are an integral part. As in many other areas, ancient Greece played an important role in the development of ideas about nature and the relationship of humans to nature. It was in ancient Greece that the very concept of nature was born. Initially, this concept referred to the mere process of realization, genesis, emergence, growing of a thing, but it gradually evolved to mean the invisible power that carries out this process, and finally a personified ideal being, the mother of all things. Two different attitudes developed in ancient Greece toward this "Mother Nature," which philosopher Pierre Hadot (2004) called "Orphic" and "Promethean." The "Orphic" attitude—named after Orpheus, the poet and musician hero of Greek mythology—consists in revealing the secrets of nature by sticking to perception, without the help of instruments, and using the resources of philosophical and poetic discourse or those of pictorial art. By contrast, the Promethean attitude—named after Prometheus, the titan who stole the sacred fire of Olympus from the gods to give it to humankind, in Greek mythology—consists of using technical processes to wrest its secrets from nature in order to exploit it. It is this Promethean attitude that has largely given rise to modern civilization and the global rise of science and industry. But it was already germinating in Greek civilization, where the manufacturing arts predominated. The ancient Greeks already conceived the making of objects as an act of creation, whereby a human subject gave form to inert matter. Similarly, they conceived of agriculture as a violent enterprise by which, year after year, humans raped the earth and plucked from its bowels the fruits that the gods had hidden from them (Pelluchon 2011). The myth of Prometheus symbolically testifies to this attitude of defiance of the gods by humans. Although Greek philosophy still conceived of humans as part of nature, it already contained the seeds of the divorce between humans and nature that modern civilization would later consummate.

It was the emergence of Christianity and the other monotheistic religions, however, that constituted the second crucial stage, after the Neolithic transition, in the process of separation of humans from nature. It is probably no coincidence that the great monotheistic religions—Judaism, Christianity, and Islam—all originated in the Middle East, which was the focus of the Neolithic revolution. Judaism and then Christianity and Islam are based on a set of stories, legends, and myths that were passed down orally, from generation to generation, for millennia before being written down on parchment or papyrus (Mason 2005). In particular, the myths of Genesis existed long before they were codified by the scribes of the Torah, the Bible, and the Koran. One cannot fail to be struck by the similarity between the biblical account of the Fall of Man and the actual history of the Neolithic transition in the Middle East, which, as we now know, represented a particularly painful change in people's way of life. It would not be surprising if such a profound upheaval could have given rise to such an ancient and powerful myth as that of the Fall of Man. Genesis could thus be a mythified representation of the Neolithic transition that took place thousands of years earlier in the same region.

In any case, Judaism and Christianity played a major historical role in asserting the uniqueness and specificity of the human species in the face of a nature that became external to it, thus prolonging in the realm of ideas the up-heaval of real living conditions brought about by the Neolithic revolution. By proclaiming the existence of a single God, the monotheistic religions were in fact proclaiming, in contrast to the proliferation of peoples and gods that had previously existed, the existence of a single, universal humankind created in his image. This proclamation undoubtedly played a major role in the gradual unification of humankind that took place during the following two millennia. It also undermined the moral acceptance of slavery that prevailed at that time, thereby contributing to its demise. But the monotheistic religions elevated humankind only to lower the rest of nature by comparison. They endowed the new universal humans created in the image of God with a historical mission that radically distinguished them from all other living beings. Immediately after creating man and woman, God said to them, according to the Bible: "Be fruitful and increase in number; fill the earth and subdue it. Rule over the fish in the sea and the birds in the sky and over every living creature that moves on the ground" (Genesis 1:28). The contrast with the hunter-gatherer worldview is striking: while hunter-gatherers were content to live carefree on the land and saw themselves as the equals of the animals, the humankind of the Bible is created to subdue the earth and rule over all other creatures. Christ himself can be seen as a humanized version of the mythical Greek Prometheus: just as the Greek titan was condemned and tortured by the pagan gods for bringing divine fire to mankind, so Christ was condemned and tortured by the author-ities and the mob of the time because he brought divine truth to mankind. The mission of this new Prometheus was based entirely on the division of the world into a material world here below and a transcendent spiritual world above (Flahaut 2008).

Judaism and Christianity not only separated humans from nature, they asserted humans' *duty* to separate themselves from nature and exercise do-minion over it. Thus, the great monotheistic religions can be seen as a form of ideological justification and apotheosis of the Neolithic transition and the evils that accompanied it. Agriculture is no longer a choice, but a moral obligation: all virgin land *must* be conquered, cultivated, and made produc-tive for humans; it is *immoral* to leave land uncultivated. It is also immoral to leave humans uncultivated. Indeed, it is humans' task to dominate nature outside them, but also nature inside them. Hence the insistence of mono-theistic religions on the duty of humans to tame their impulses and needs, which are manifestations of their natural, animal heritage. The divorce be-tween humans and nature that they proclaim is thus accompanied by a di-vorce within humans themselves—between, on the one hand, their body and its animal needs, and, on the other, their mind and its ideals, considered as specifically human. The struggle between Good and Evil is just another way of translating the inner conflict that the Christian religion anchors into the heart of human beings, between their spiritual and ideal dimension, turned toward God and supposed to be specifically human, and their bodily and material dimension, anchored in nature and deeply animal.

The rise of Christianity is often associated in popular imagery with a period of regression, both in terms of intellectual and material production, hence the term "Middle Ages." In reality, this was not the case. It is true that freedom of thought regressed during this period, but the rise of Christianity did favor the domination of humans over nature, which it contained as a precept. In particular, the Middle Ages contributed to the development of agriculture, which had begun in the slavery societies of antiquity after the Neolithic transition. New agricultural techniques, such as the three-year crop rotation, the replacement of oxen by horses, and the invention of the wheel and moldboard plough, considerably increased agricultural productivity during this period. At the same time, the introduction of wind and water mills provided a new source of energy that made it possible to increase the yield of grain milling; the improvement of cartography and navigation instruments paved the way for the great maritime discoveries; and the invention of mechanical clocks profoundly changed humans' relationship with time. All of these technical innovations laid the foundations for a considerable demographic and economic expansion, the growth of cities, and finally the rise of capitalism.

The final stage of the long process leading to the divorce between humans and nature comes with "modernity," a period following the Renaissance in which both the modern Western worldview and capitalism as an economic and social system were formed, both reinforcing each other in a spiral of hegemonic expansion that gradually dissolved all ancient forms of thought and social organization around the world. Modernity is often presented as a break with the Middle Ages and its religious worldview. But the violent ideological oppositions between "Ancients" and "Moderns" in the centuries following the Renaissance were mainly about the place of religion in civil society, in the state, or in science; they were not about substance, which was common. Similarly, the wars of religion that bloodied Europe in the sixteenth and seventeenth centuries did not mean that two antagonistic worldviews clashed; Catholicism and Protestantism were just two versions of the same worldview, and even of the same religion.

If we take a step back from the conflicts of the time, it appears that modernity and its economic manifestation, capitalism, simply stripped Christianity of its idealistic trappings and anchored it in the earthly world and in the daily reality of human beings. Christianity endowed humans with a historical mission;

modernity and capitalism undertook to fulfill it. Philosophers Francis Bacon and René Descartes, who are often considered the founding fathers of modern science, are very explicit on this point. In his masterpiece, *Novum Organum*, Bacon (2016 [1620], 32, 41, 35, 80) presents science as the means of fulfilling the mission entrusted to humans by God, which is to increase human power by subjugating nature: "But if the matter be truly considered, natural philosophy is, after the word of God, at once the surest medicine against superstition and the most approved nourishment for faith, and therefore she is rightly given to religion as her most faithful handmaid, since the one displays the will of God, the other his power." "My purpose (. . .) is to try whether we cannot in very fact lay more firmly the foundations and extend more widely the limits of the power and greatness of man." "For I do not run off like a child after golden apples, but stake all on the victory of art over nature in the race." All means are good to achieve this end, including violence, for "the secrets of nature reveal themselves more readily under the vexations of art than when they go their own way." Thus, modern science was conceived, not as a disinterested search for truth—as it is often presented, and sincerely experienced by many scientists—but as a conscious and systematic undertaking to conquer and subjugate nature. "For since our main object is to make nature serve the business and conveniences of man, it is altogether agreeable to that object that the works which are already in man's power should (like so many provinces formerly occupied and subdued) be noted and enumerated."

While Bacon displayed in a particularly transparent way the Promethean will to power and domination hidden in modern science and industry, Descartes provided its philosophical foundations. Descartes is rightly regarded as the genuine theorist of modern rationalism. In his famous statement, "I think, therefore I am," Descartes principally asserts the separation between the immaterial soul that defines the human thinking subject, on the one hand, and the material body that defines the objects of nature, on the other. Descartes strips nature of all enchantment and reduces it to a set of objects obeying purely mechanical laws, created by God and which humans can dispose of as they please. Descartes no longer even speaks of nature: for him, the world is one and homogeneous, an infinite space made up of pure expanse in which inert matter moves under a divine impulse. This neutral, soulless, meaningless world is radically different from the thought that conceives it, an immaterial attribute with which God has endowed humankind.

The divorce between humans and nature is thus consummated on a philosophical level. Humans as thinking subjects no longer have anything in common with the material world and the rest of life. They must even distrust their senses, which are in contact with matter and can therefore be corrupted by it and mislead them. Only rational thought allows humans to be human, that is, akin to God: "I am therefore, precisely speaking, only a thinking thing, that is, a mind, understanding, or reason." "And thus I very clearly see that the certitude and truth of all science depends on the knowledge alone of the true God" (Descartes 2021 [1641], 19, 53).

If, according to Descartes, living beings are only soulless machines, some of his disciples did not fail to deduce, quite logically, that they are insensitive and therefore incapable of suffering. Descartes's philosophy thus came to justify not only human domination over nature, but also the unrestrained violence that this domination has sometimes taken. Vivisection has been practiced without remorse for centuries by biologists on the pretext that animals cannot truly suffer, whatever the apparent manifestations of their suffering. Nicolas Malebranche, a seventeenth-century philosopher, mocked dogs' expressions of pain as follows: "What groans, what howls, what sensitive marks of a very cruel pain! All this is only a game of machines" (cited in de Fontenay 1998, 296).

The thesis of automatism was so much at odds with the reality experienced by his contemporaries that Descartes proposed a fiction to destroy this illusion inherited from childhood, which would have us believe that animals act according to an inner principle similar to ours and that they possess a soul, feelings, and passions. One must imagine, he writes, "a child who has never seen animals, but only men, and who has further studied mechanics to the point of helping to make automatons that imitate the figure and movements of a man, a horse, a dog, automatons that seem to come and go, to breathe and even to speak" (cited in de Fontenay 1998, 284). In order to defend the validity of his thesis, Descartes therefore appeals to a utopian world in which humans would be cut off from all contact with nature. In this utopian world, humans would have no choice but to get used to the idea that the world consists only of humans and automatons, since that is indeed all it would contain. This dream of a fully humanized and automated world could be seen as an inconsistency of Descartes's thesis, since he has to imagine a world in the image of his thesis to demonstrate its validity, which is an obvious example of circular reasoning. But this apparent logical inconsistency does not affect the basis of his philosophical thinking. For the revolution of reason he advocated cannot be understood as a simple attempt to explain the world as it presents itself to us. On the contrary, Descartes posits his method as a new requirement that the human will must assume and carry through (Janicaud 2005). The utopian dream he proposes as fiction is, in fact, the program of modernity: the creation of a mechanical and automated world at the service of a new human species, freed from all material constraints and thus effectively reduced to a thinking soul, a modern version of the Promethean myth exalting human omnipotence over nature and its pagan gods.

Capitalism, which took off at the same time, is the economic form that this program takes. As Marx demonstrated in the nineteenth century, capital is not a thing or a set of things, but first and foremost a process of production and reproduction, on an ever-expanding scale, of exchange value. Now, exchange value is a social attribute that expresses the equivalence of a traded commodity to the universal commodity, which is money. Thus, it is a relation of equivalence that reduces naturally heterogeneous commodities to an abstract, homogeneous social quantity. This relation of equivalence erases not only the natural qualities of the things exchanged, but also the non-market values that people may attribute to them. In other words, market exchange performs the same operation in the relations between humans as Descartes's reason does in thought: it reduces the infinite richness of nature and of humans' relations to nature to a pure abstract expanse in a homogeneous monetary space. The word "reason" comes from the Latin *ratio*, which means ratio, calculation. It thus appears clearly that capitalism is the economic expression of modern rationality and, conversely, that modern rationality is the philosophical expression of capitalism. The industrial revolution has enabled capitalism to realize, to a large extent, what Bacon and Descartes predicted, namely an automated world, in which nature and humans themselves tend more and more to be replaced by machines. Machines are not necessarily opposed to humans and nature, but mechanization as a program for the subjugation of nature is incompatible with the flourishing of nature and of humans within it.

While Marx provided a remarkable analysis of capital, its historical genesis, its foundations and its contradictions, he did not succeed in freeing himself from the worldview of his time, and in particular from the Promethean myth exalting human omnipotence over nature. Although Marx criticized modern rationalism and took a great interest in the ecological problems of his time, such as soil fertility depletion and deforestation (Foster 2000), it is no coincidence that he concluded the foreword to his doctoral thesis with the sentence: "Prometheus is the most eminent saint and martyr in the philosophical calendar." Throughout his life, Marx advocated the emancipation of humans from all forms of slavery, both spiritual and material. In doing so, he placed great emphasis on everything that makes human beings active subjects in the transformation of the world and of their own transformation. In particular, in the economic sphere, he viewed production and labor as social processes by which humans assert themselves as active subjects in the face of passive matter which they use to produce an object. On his view, although humans are part of nature defined as the totality of reality, the worker faces and opposes the rest of nature as a creative agent in the process of labor. Thus, labor breaks the primitive unity of humans and nature to assert the unique social and historical character of the human species (Schmidt 1971). By exalting the creative force of humans confronting nature in the process of production, Marx and Marxism are in line with the continuity of the Promethean program of human mastery of nature outlined in ancient Greece, developed by Judaism and Christianity, and completed by modernity and capitalism.

Although Marx announced the end of capitalism, his vision of the new society that would emerge from its flanks was entirely oriented toward human emancipation. Beyond the realm of necessity, he wrote, "begins that development of human energy which is an end in itself, the true realm of freedom." Marx spoke of the development of human energy, not of a new relationship between humans and nature; he spoke of the end of the exploitation of humans by humans, not of the end of the exploitation of nature by humans. Thus, despite Marx's interest in the ecological problems of his time, the theoretical framework he developed does not overcome the separation between humans and nature that is generating the current global ecological crisis.

Despite the growing awareness of the extent of this global ecological crisis in recent decades, the world in which we live is still profoundly shaped by a way of thinking and social relations whose hidden or avowed aim is to tear humans away from nature in order to bring them closer to the image of the God they have created for themselves—an immaterial spirit that dominates and shapes the material world in which they are immersed. Certainly, reason is no longer invoked today with the same enthusiasm and blindness as it was a few centuries ago, which has led some sociologists to proclaim the end of modernity and the emergence of a new historical phase of "postmodernity" in contemporary Western societies. The rationalist edifice that serves as the ideological underpinning of modern society as a whole is beginning to crack on all sides, as it demonstrates more and more its limitations in the face of the expression of irrational forces at work in the world, both human and non-human. But despite its superficial cracks, this edifice based on the separation between humans and nature remains largely intact. In particular, it permeates the entire world economy, in which the power of exchange value, money, and capital is asserting itself with ever-increasing force. Not only are we not yet out of modernity, but, in some respects, we are in it more than ever. The global contemporary human society looks more and more like the utopia imagined by Bacon and Descartes.

The only currents that sought to emancipate themselves from modernity and proposed a different vision of the relationship between humans and nature toward the end of the twentieth century were those of environmental ethics and "deep ecology." The former became an academic discipline, mainly in Anglo-Saxon countries, while the latter gave rise to the so-called green movements and parties around the world. These two currents have produced a multifaceted critique of the separation between humans and nature that underlies modernity. Unfortunately, this critique has largely failed so far to produce an alternative forward-looking worldview. As a result, the ideologues of modernity have not failed to portray deep ecology as a romantic aspiration to an illusory "return to nature" belonging to a bygone era of history, or even human prehistory. While deep ecology has had a significant impact in the most economically developed countries, as evidenced by the emergence of political movements inspired by it, the critique of modernity that underpins it remains poorly known, even though it is far more fertile than the political ideas that emerged from it. The following chapters will echo some aspects of this critique.

The destruction of nature is not unique to modernity, since examples can be found at least from the Neolithic, in particular during the human colonization of new territories. But monotheistic religions and modernity have made the submission of nature a foundation of their social project. We are therefore faced with a historical situation that has nothing in common with that of Paleolithic or Neolithic societies. Today, it is the very worldview that has been passed on to us since our earliest childhood that stands in the way of an evolution of contemporary society to a more respectful attitude toward our natural environment. Only by uprooting the most deeply rooted beliefs in this worldview can we give ourselves the means to overcome the current global ecological crisis. This is what the following chapters will seek to do.

3
Subject and object: The mirror of modernity

The notions of subject and object are so deeply rooted in the modern world-view that we take them for granted. Is it not obvious that in every thought and action there is a subject who thinks or acts and an object to which that thought or action relates? Does not every sentence we utter, from the time we learn to speak to the time we die, bear witness to the fact that every action involves a subject who acts and an object on which it is carried out? And then, what do these grammatical considerations have to do with the relationship between humans and nature that concerns us here?

In truth, the subject-object couple we are so familiar with is a pure product of modernity and has everything to do with modern society's relationship with nature. Subject and object are not just grammatical constructs; they are key concepts that crystallize the modern worldview and, in particular, the domination of humans over nature. To take an interest in this couple, which is omnipresent in our daily lives, is to question one of the most elementary and pervasive collective fictions that govern our relationship to the world without us even noticing.

What is it about? The notion of object is quite simple to grasp. The term "object" comes from the Latin word *objectum*, which means "that which is thrown in front." What is thus "thrown in front" is anything to which a thought or action is directed. By contrast, the notion of subject is at first sight much more heterogeneous, and even contradictory. The term "subject" comes from the Latin word *subjectum*, which literally means "that which is thrown under." But what is "thrown under" has come to mean three different things in the course of pre-modern history. On the one hand, there is the political subject: subjects are "thrown under" their ruler, that is, they are subject to their ruler's domination and power. The political subject is thus far from being the one who exercises action; on the contrary, he is the one who undergoes the power of his sovereign, the only real holder of political action. On the other hand, there is the philosophical subject: for Aristotle, subject designates what is "underlying" something, that is, its substance or what makes this something what it is. The philosophical subject is therefore not necessarily linked to action

Nature That Makes Us Human. Michel Loreau, Oxford University Press. © Oxford University Press 2023.
DOI: 10.1093/oso/9780197628430.003.0004

either: it is closer to "being" than to "doing" or "having." Finally, there is the grammatical subject, the one who performs the action described by the verb and who seems, at first sight, closer to the subject as we understand it today. Although there were links between these different forms of subject, there was no real coherence between them at the dawn of modernity.

Modern philosophical discourse unified this fragmented notion of the subject by giving it a new dimension. No one expresses this conceptual upheaval better than Descartes. Heidegger (1971) showed how Descartes's famous formula *Cogito, ergo sum* ("I think, therefore I am") is the affirmation of a new vision of humankind and of its place in the world. In ancient metaphysics, all beings were conceived as subjects, that is, as things that determine themselves by what they are. Thus, stones, plants, and animals were no less subjects than humans. Descartes, on the other hand, proposed a radically different conception. *Cogito, ergo sum* does not only say that I think, that I am, or that my existence results from my thinking. It states that I am insofar as I represent myself in thought, that my act of representing in thought decides the presence of all things represented. Thus, humankind becomes the foundation of the representation of everything, and therefore the only true subject. Consequently, all non-human beings become objects for this subject. Humankind as subject gives the measure of all things, becomes the center of the world represented in its totality and is no longer constrained by any limits.

In fact, for Descartes, humankind no longer truly belongs to the world, for the latter is now split into two parts that no longer have anything in common: on the one hand, inert matter, pure expanse that obeys the implacable laws of mechanics; on the other, the human thinking soul or spirit, which derives its freedom from divine grace. It is therefore logical that the philosophical subject ceases to be the substance of everything and becomes the thinking substance of the human soul, which faces the extended substance of the material world. And since this thinking substance is proper to humans, the only true subject is now the thinking human being capable of affirming, "I think, therefore I am," in the face of the inanimate world. We can thus see that this new credo provides the philosophical foundation for human domination over nature (Derrida 2006). Humankind as subject conquers the world through thought before conquering it through physical action.

In the political sphere, this philosophical credo would soon be translated into the emancipation of the political subject, who claims the primacy of individual action in the face of an objective world that has become inert. Similarly, in the economic sphere, the individual would be elevated to the status of economic subject, who affirms the freedom of individual action in the face of the market. Thus, the Cartesian thinking subject would henceforth merge with the emancipated political and economic subject to become the modern free individual as we know him/her today.

For all that, the new subject created by modernity has not freed itself from the constraints that made it a political subject, that is, an individual subject to a higher authority. For the Cartesian thinking subjects assert their freedom and power only in relation to the material world, but derive their freedom and power from the divine will, to which they remain subject. This submission, however, changes form: it ceases to be blind obedience and becomes freely consented obligation. In place of the certainty of divine salvation which gives the measure of all truth in Christianity, the modern human subject posits the certainty of being oneself through the exercise of rational thought. But this new certainty is based on new obligations that enable one to guarantee one's approach, in particular the obligation to submit to reason (Heidegger 1971).

Just as the new philosophical subjects are only free insofar as they submit to the philosophical order based on the use of rational thought that consecrates them as subjects, the new political and economic subjects are only free insofar as they submit to the political and economic order that guarantees their freedom of action in the world, that is, to the authority of the state and the market. Thus, the unification of the subject effected by modernity is based on a new combination of freedom and subjugation, in which subjugation is hidden in the very exercise of freedom. The motto of the French revolution, "Liberty, Equality, Fraternity," which exalts the free individual, cannot be conceived without its counterpart of obligations that subject these individuals to a shared framework for thinking and living together, which could be summarized by the formula: "Reason, State, Market."

The inception of the modern subject marks an important turning point in the history of human thought because it fundamentally redefines humans' place in the world. In particular, it consummates their separation from the rest of nature. Modern human subjects see themselves as alone in the world, alone with their consciousness, their thought, and their free will inherited from an immaterial divinity. The rest of the world faces them as an inanimate, purely material, mechanical nature, governed by a mixture of chance and necessity. Humankind has nothing more to expect from this mechanical nature. The only thing it can do with it is to dominate it and exploit it to the best of its ability to increase its power. "God is dead," philosopher Friedrich Nietzsche proclaimed, with a mixture of despair and exaltation; consequently, the only motive for human action, according to him, is the will to power. But is this will to power that is so strongly asserted in modernity anything other than the mission that the Christian God entrusted to humans when he enjoined them to "fill the earth and subdue it"? In this sense, God is not dead: he has transmuted himself into Reason, the modern expression of the Promethean will to power.

As to what deeper meaning this mission of increasing human power might have, this question remains forever unanswered. Christianity, like the other monotheistic religions, gives a semblance of meaning to this mission by presenting it as a manifestation of the divine will, but this divine will is arbitrary, as it comes from an all-powerful God located outside the real world in which humans live. Modernity in its developed form, stripped of its religious trappings, eliminates any trace of meaning in nature since the latter is reduced to a mechanical material world, abandoned to the blind forces of chance and necessity. As Harari (2015, 199) puts it, "Modernity is a deal. All of us sign up to this deal on the day we are born, and it regulates our lives until the day we die. (. . .) The entire contract can be summarised in a single phrase: humans agree to give up meaning in exchange for power." Where I believe Harari goes wrong, however, is that we do not accept this deal on the day we are born. Young children do not spontaneously seek power; instead, they spontaneously tend to find meaning everywhere in the world around them. It takes a long process of persuasion and indoctrination, which takes place through what we call education, for children to eventually erase all traces of meaning in the world and sign the contract of modernity, often reluctantly. Giving up meaning in exchange for power not only often plunges modern humans into deep philosophical and psychological distress, as is widely recognized today, it also has a powerful hypnotic power: it makes them forget what they are and mobilizes their energy and intellectual capacities toward a goal that does not truly belong to them. The modern free subject is the product of a gigantic collective fiction—perhaps the most powerful in history—and the plaything of social forces that are completely beyond their control.

This state of affairs naturally has serious consequences for the relationship of the modern subject to nature. For the spiritual father of modernity, Descartes, humans are radically different from nature not by their material body, but by their immaterial thinking soul, which links them to God. The divine character of this thinking soul, however, is not essential, and indeed disappears in the more materialistic version of modernity, which eliminates all traces of the divine in the real world. What is essential is that, through their thinking soul, humans assert themselves as the sole subjects in relation to the rest of nature. What Descartes affirms with his "I think, therefore I am" is, in fact, "I think, therefore I am subject, and the rest of the world is object." This radical separation of subject and object is accompanied by an equally radical transformation of their relationship: since the subject is conceived as the underlying foundation of all representation of the material world, it is now situated above the world of objects, which it has the mission and essence of dominating. It is therefore very logical that Descartes would assert that the new status of humans as modern subjects can make them "like masters and possessors of nature." This long-debated phrase merely translates into concrete terms the preeminence of subject over object, of spirit over matter.

The new contours of subject and object defined by modernity go hand in hand with the preeminence of labor and production in the modern capitalist economy. Marx showed how, in the modern labor process, humans face raw material as active subjects that give it form and use. But what Marx presented as an immutable law of human labor is only a relatively recent invention, which dates back to the Neolithic transition at the most and which capitalism has generalized and perfected. As anthropologist Philippe Descola (2005, 440) rightly notes, "Marx's position is indicative of a more general tendency in modern thought to privilege production as the determinant of the material conditions of social life, as the main way in which humans transform nature and, in so doing, transform themselves." It is, in a way, the transposition of the Promethean myth into the economy.

In the hunter-gatherer societies of the Paleolithic or today, the very concepts of production and labor are meaningless because they are incompatible with the worldview of these societies. Even in the traditional Chinese civilization, which developed within a highly agricultural and state-controlled society, "the world is not produced by the intervention of an actor with a design and a will, it is the result of its internal propensities alone, which manifest themselves spontaneously in a permanent flow of transformations. One can measure the chasm that separates this self-regulated process from the heroic model of creation as it developed in the West as an unquestioned evidence under the double aegis of the biblical tradition and Greek thought. The idea of production as the imposition of form on inert matter is only an attenuated expression of this scheme of action, which rests on two interdependent premises: the preponderance of an individualized intentional agent as the cause of the advent of beings and things, and the radical difference in ontological status between the creator and what he produces. According to the creation-production paradigm, the subject is autonomous and his intervention in the world reflects his personal characteristics: whether he is a god, a demiurge or a mere mortal, he produces his work from a pre-established plan and in function of a certain finality." In contrast, "far from being understood as the production-creation of a new thing from inanimate matter informed by the art and project of an autonomous agent, the work of the Wayana basket-maker is conceived as what makes possible a true metamorphosis, that is to say, the change of an entity already existing as a subject and which retains all or part of its attributes in the operation" (Descola 2005, 442, 444–445).

The subject-object duality perfected by modernity is particularly pernicious because, once internalized as a norm in consciousness, it seems to be self-evident. Thus, it determines, without our realizing it, all our thoughts and social behaviors, particularly with regard to nature. The concept of nature itself merely reflects this duality: "nature" refers precisely to everything that does not belong to thinking, acting human beings and their free will, in short, to the modern subject (Evernden 1992). Nature is therefore the realm of objects, of things that do not think and do not act, of inert matter. Of course, nature includes living beings that move, feel, and react, but, as we saw in the first chapter, these living beings are supposed to have no consciousness or autonomous action; they are merely cogs in an immense universal mechanism subject to the blind laws of chance and necessity, and thus they do not truly act. The human body itself is part of this nature because it is made of matter. Only their thought makes humans subjects and radically separates them from the material world that makes nature.

Of course, many things have changed since Descartes, including regarding the subject-object duality. Few scientists today still believe that thought so radically separates humans from nature. Some tend toward an integral materialism, which dissolves human consciousness in the mechanical material world, others toward a divinization of nature which brings it closer to humans. Upon closer inspection, however, this duality continues to govern our way of thinking and acting to a very large extent. Without fully recognizing its origins, contours, and consequences, we will not be able to profoundly change our relationship with nature and with ourselves. We will remain prisoners of illusions inherited from hundreds, even thousands of years of history without even realizing it, and we will only be touching the surface of the issue.

For example, despite the many changes in the modern worldview that have taken place since Descartes, few scientists or philosophers would be prepared to put animism and rationalism on equal footing and to assert that the non-human world is animate and composed of subjects, just like the human world. Whatever their differences and their materialistic or spiritualist leanings, the vast majority of contemporary scientists and philosophers consider that animism is a worldview inherited from primitive societies, forever superseded by the modern worldview. I have no intention to defend one worldview against the other, but I cannot see why animism would be a more detestable or retrograde collective fiction than rationalism. As a matter of fact, animism is much closer to our innate relationship to the world, which explains why, in childhood, people tend to spontaneously adopt an animistic worldview and why, in adolescence and adulthood, it takes intense and re-peated efforts to inculcate them with a rationalist worldview. Rationalism is a collective fiction that fulfills the stated aim of modernity, to elevate humans out of and above their nature. Therefore, this collective fiction can only be anchored in individual consciousness through a long and difficult social pro-cess called education. This is in particular the role of the school, an institu-tion that ensures an efficient and organized reproduction of the collective fictions on which the whole of modern society is based and without which it cannot function.

Why do most people—at least in economically and ideologically "devel-oped" countries—take it for granted that the animals, plants, rocks, moun-tains, and rivers around us do not have a soul? Admittedly, the notion of soul has become extremely blurred, to the extent that many contemporary scientists and philosophers consider soul to be a useless concept that does not exist in animals, plants, or even humans, let alone, of course, in rocks, mountains, and rivers. But getting rid of the concept of soul does not solve the problem. For many of the same scientists and philosophers continue to believe that humans are nevertheless an exception because they possess self-awareness, thought, language, free will, or any of the other intellectual attributes that we examined in the first chapter and that are supposed to make them the only true subjects. In other words, the immaterial soul has been replaced by other attributes, and the divine origin of these attributes has sometimes disappeared among athe-istic materialists, but *Homo sapiens* is still a creature apart in the universe. And yet, as we have seen in the first chapter, this firmly held belief is not based on any serious scientific evidence.

This belief is also in direct opposition to our spontaneous perception. As philosopher David Abram (1996, 130) rightly reminds us, "direct, prereflexive perception is inherently synaesthetic, participatory, and animistic, disclosing the things and elements that surround us not as inert objects but as expres-sive subjects, entities, powers, potencies." You do not need to believe in ani-mism to be convinced of this. If you free your perception from the grip of the intellectual constructs with which your consciousness is encumbered and allow your attention to take in what comes to you without any expectation, without any prejudice, without any thought, even if only for a moment, you will be struck by the truth of these words. It is only afterward, when the pro-cess of intellectual reflection starts again, that this obvious and immediate truth tends to fade away, sometimes very quickly. Experience it yourself if you have not already!

No wonder our tribal ancestors considered that non-human animals, plants, rivers, and mountains spoke to them. In a way, they do speak to us, but not in the words we are familiar with. They speak to us by their physical presence, by the air they move, by their smell, by their taste, by their texture to the touch, by their look, by the sounds they emit. They also move, although sometimes very slowly, and above all they act and transform our environment, even when they appear to be immobile. Their apparent immobility results from an optical illusion that makes us see what moves more slowly than us as being devoid of movement. Rivers and mountains have shaped the landscapes in which we move today over millions and hundreds of millions of years, exerting considerable forces step by step. If they exert a force, they act. And if they act, they are active subjects. The ultimate paradox of modernity is that it has sought to remove the character of nature as an active agent in order to be able to exploit it without restriction, but in so doing it has set in motion a movement of natural agents so powerful that this movement threatens its very existence! One need only think of the disastrous consequences that current climate change and biodiversity loss are likely to have for future human generations to be convinced of this.

If we accept the undeniable fact that the entities that make up this abstract "nature" we have conveniently dismissed from us are active agents of their own destiny and ours, then there can be no justification for stripping them of subject status and reducing them to inert objects (Evernden 1993). Nature is composed of a multitude of entities that interact with each other in extraordinarily complex ways, and we are part of them. These entities are neither subjects nor objects in themselves: they simply exist, live, and transform each other, and thus participate in the cosmic flows of matter, energy, and life. If we wish to use the concepts of subject and object, then it must be clear that they are simply the translation, at a given moment, of a relationship between entities placed on an equal footing. Each entity may be both subject and object to other entities, but this relationship is constantly changing and exchanging, and therefore says nothing about the nature of the entities in question. Abram (1996, 67) illustrates this idea with a telling example:

> contemporary discourse easily avoids the possibility that both the perceiving being and the perceived being are *of the same stuff*, that the perceiver and the perceived are interdependent, and in some sense even reversible aspects of a common, animate element, or Flesh, that is *at once both sensible and sensitive*. We readily experience this paradox in relation to other persons; this stranger who stands before me and is an object for my gaze suddenly opens his mouth and speaks to me, forcing me to acknowledge that he is a sentient subject like myself, and that I, too, am an object for his gaze. Each of us, in relation to the other, is both subject and object, sensible and sentient. Why, then, might this not also be the case in relation

to another, nonhuman entity—a mountain lion, for instance, that I unexpectedly encounter in the northern forest?

Relativizing the concepts of subject and object as mere roles in an ever-changing relationship between entities does not necessarily imply any form of animism. The opposition between rationalism and animism is another straitjacket created by modernity that keeps us trapped in its dead-end world-view. And to get out of this straitjacket, understanding and appreciating the perspectives provided by other worldviews is extremely valuable. As a matter of fact, the animistic worldview developed by many hunter-gatherer societies is so far removed from the modern worldview that the very concepts we use to describe it are a constant source of confusion and misunderstanding. A good example is the term "spirit" used by the Amerindian peoples. Native Americans do not see the "spirit" as a mysterious power that resides inside their heads, as Christianity and modernity conceive it, but as a quality within which they find themselves, along with other animals, plants, rivers, and mountains. This universal quality can even take concrete forms, such as air and wind. "For the Navajo, then, the Air—particularly in its capacity to provide awareness, thought and speech—has properties that European, alphabetic civilization has traditionally ascribed to an interior, individual 'mind,' or 'psyche.' Yet by attributing these powers to the Air, and by insisting that the 'Winds within us' are thoroughly continuous with the Wind at large—with the invisible medium in which we are immersed—the Navajo elders are suggesting that that which we call the 'mind' *is not ours*, is not a human possession. Rather, mind as Wind is a property of the encompassing world, in which humans—like all other beings—participate" (Abram 1996, 237).

The identification of mind with wind is not as foreign to Western civilization as one might think. Indeed, the term "psyche" derives directly from the ancient Greek word *psychê*, which signified not only "soul" or "mind," but also "breath" or "gust of wind." The word "spirit" has the same origin; it derives from the Latin word *spiritus*, which signified "air," "wind," or "breath." The Latin word for "soul," *anima*—from which the words "animal," "animation," and "animism" are derived—also initially signified "air" and "breath." This shows that "*awareness, far from being experienced as a quality that distinguishes humans from the rest of nature, was originally felt as that which invisibly joined human beings to the other animals and to the plants, to the forests and to the mountains*" (Abram 1996, 238).

All the factors that contributed to the emergence of the modern fiction that humans are radically different from nature—including the emergence of a sedentary agricultural lifestyle, of hierarchical social structures, and of alphabetic writing—deeply altered the nature of humans' conscious experience. Human consciousness, which was originally experienced as a universal breath of life passing through human beings, was gradually transformed into an experience specific to each individual, left to his or her own devices to deal with the vagaries of his or her own production or of the market, the laws imposed by an external authority, and the written texts that transmit them. It lost the unity and continuity with the natural world that it had originally to become the manifestation of an individual interiority, a private "soul" separated from the other "souls" around it and from the surrounding Earth. The original cosmic "spirit" was thus transfigured into a new, shrunken, and fragmented "mind" that is the private property of each individual human being.

As the "spirit" became increasingly narrow and internalized in individuals, it disappeared from the rest of the world. Monotheistic religions and modernity completed this compartmentalization by banishing spirit, mind, and consciousness from all non-human entities. With Descartes, even animals officially ceased to be endowed with a "soul," an animation of their own, to become purely mechanical devices. The whole of nature has thus become a gigantic machine, made of inert matter. Now, by definition, inert matter cannot move by itself. Since the material world is obviously full of all kinds of movements, some external force had to be exerted on the universe by an external active agent to explain these movements. For Newton and his contemporaries, the ultimate source of motion in the universe was God, a pure spirit that imposes the laws of nature on inert matter (Shapin & Schaffer 1985; Ellis 2002). Modern physics has now replaced the divine command with an equally mysterious initial impulse called the "big bang," which conveniently saves the appearances of the modern worldview, in which present-day matter, as we know it in everyday life, continues to be regarded as inherently inert for all practical purposes.

Thus, a double movement has taken place in Western civilization: on the one hand, a movement of over-animation of some of the actors in the theater of life—specifically humans, endowed with admirable capacities for action and thought—and, on the other hand, a movement of de-animation of the other actors—all non-human entities, reduced to inert material objects constituting the scenery for human action. As sociologist and philosopher Bruno Latour (2015, 94–95) notes, "although the official philosophy of science takes the second de-animation movement as the only important and rational one, the opposite is true: animation is the essential phenomenon; and it is de-animation that is a superficial, auxiliary, polemical and often apologetic phenomenon. One of the great enigmas of Western history is not that 'there are still people naive enough to believe in animism,' but the rather naive belief that many people still have in a supposedly de-animated 'material world.'"

The world around us is inhabited and animated by an innumerable multitude of actors who, like us and with us, participate in the great history of the evolution of the universe and of life. To fix one of these actors in a category of "acting subject" and all the others in a category of "inert objects" is tantamount to taking a snapshot of a complex, highly dynamic world from the point of view of a single actor and declaring that this snapshot is the only possible representation of that world. I guess nobody would have the audacity and stupidity to claim that the photo he or she took of such and such a person on such and such a day in such and such a place demonstrates that he or she is the only acting subject and that the person photographed is obviously an inert object since he or she is not moving. Yet this is exactly what modernity does when it decrees that only humans are acting subjects and the rest of the world is made of inert objects. Modernity has deliberately frozen the rest of nature as an inert background to glorify the transformative role of human thought and action. Reanimating the non-human world, recovering the soul, the consciousness, the spirit that humans have arrogated to themselves in the world around us is a necessary condition for recovering our place in the cosmos, in life, in evolution—and thus for recovering ourselves as part of this world.

4

Matter and spirit: The great illusion

Behind the subject-object duality characteristic of the modern worldview lies another, even more fundamental duality, which did not fail to surface in the previous chapter: the duality between matter and spirit. Descartes defined the modern thinking subject precisely by the fact that it possesses an immaterial soul, in other words a "spirit" or "mind," while the rest of the world is made up of objects because it does not possess one. Thus, the Judeo-Christian religious tradition and then modernity came to distinguish two mutually incompatible worlds: that of spirit and mind, and that of matter and body.

> To summarize: human actions and experiences were *mental* or spontaneous outcomes of reasoning; they were performed, willingly and creatively; and they were active and productive. Physical phenomena and natural processes, by contrast, involved brute matter and were *material*: they were mechanical, repetitive, predictable effects of causes; they merely happened; and matter in itself was passive and inert. Thus the contrast between reasons and causes turned into an outright divorce, and other dichotomies—mental vs. material, actions vs. phenomena, performances vs. happenings, thoughts vs. objects, voluntary vs. mechanical, active vs. passive, creative vs. repetitive—followed easily enough. (Toulmin 1990, 108)

Behind the two seemingly opposing notions of matter and spirit, however, lies an immense confusion, which has been perpetuated for centuries. Not only does this confusion contribute to maintaining the subject-object duality in our thoughts and actions, it also prevents us from seeing clearly what is at stake in the current ecological crisis and how to resolve it, because it keeps us trapped in a pernicious opposition between materialism and spiritualism. I myself was a convinced materialist for a long time; my reflections on the origin and consequences of the current ecological crisis led me to change my mind about the nature of materialism and spiritualism. Therefore, I believe it might be useful for me to share these reflections here, in the hope that they might also be useful to others in changing their views and finding new ways to overcome the ecological crisis.

Nature That Makes Us Human. Michel Loreau, Oxford University Press. © Oxford University Press 2023.
DOI: 10.1093/oso/9780197628430.003.0005

The concepts of matter and spirit and the worldviews that are based on them—materialism on the one hand, idealism or spiritualism on the other—have been hotly debated by philosophers and scientists throughout the history of Western civilization since ancient Greece. As I am neither a philosopher nor a physicist, I will not venture into the intricacies of these debates. But to the non-specialist that I am, one thing seems striking: despite the major advances made by modern physics in the twentieth century, these advances have done little to alter the fundamental worldview characteristic of modernity. It is true that quantum physics has given rise to many questions, and sometimes even delusions, about the material and spiritual dimensions of the universe in which we live, but what has come out of it in terms of coherent philosophical worldviews? Nothing or not much that is accessible to the general public, and therefore nothing or not much that has profoundly changed our worldview. A few physicists, like David Bohm (1983), have made remarkable efforts to present a new worldview in line with the knowledge of contemporary physics, but these efforts have remained very much in the minority, even (or perhaps especially) within the scientific community, and thus have not strongly affected the thinking of ordinary people. Yet, regardless of the interpretations of some of its most misunderstood aspects—notably in quantum physics—contemporary physics provides a number of scientific elements that profoundly challenge the worldview of modernity, in particular the concepts of matter and spirit that form its foundation.

Beyond the countless forms they have taken throughout history, materialism and idealism or spiritualism are essentially opposed on one point: which, matter or spirit, has primacy over the other? The many variants of materialism have in common the belief that matter is primordial, while spirit or mind—often equated with consciousness—is supposed to be a property of matter that only appeared with the development of the brain in humans, or possibly higher vertebrates. By contrast, the many variants of idealism and spiritualism have in common the belief that it is spirit—often equated with an immaterial soul, or even God—that has primacy over matter. A key issue from the point of view of materialism is that "spirit" or "soul" are ill-defined and unobservable concepts. Therefore, their existence seems to be based on pure belief (faith, in religions), whereas the existence of matter can be empirically verified by anyone under any circumstance. This is the main reason why modern science, which bases knowledge on observable and experimentally verifiable facts, has often been invoked in support of materialism. It would seem, therefore, at first sight, that the debate is over and that modern science is only compatible with materialism.

But is this really the case? And, to begin with, what exactly is this thing we call "matter"? In physics, matter is defined as what makes up any body that has a tangible reality; concretely, matter occupies space and has mass. With the advent of atomic physics, it may have seemed for a moment that physics had solved the enigma of matter by demonstrating the universal presence of atoms, that is, literally, unbreakable elementary entities. But this illusion was short-lived; it was quickly shattered by the subsequent discovery of a veritable bestiary of subatomic elementary particles, the nature and properties of which remain a thick mystery to the average person. Worse still, there are anti-matter particles, as well as massless particles, which are therefore incompatible with the traditional definition of matter—such as the photon, the elementary particle associated with light and electromagnetic waves. Thus, modern physics teaches us that, when we look for the constituent elements of matter, we come across elements that no longer meet the very criterion that defines matter. Although this proliferation of elementary particles does not frighten physicists at all, who on the contrary find it a reason for new and exciting research, it raises very serious questions about the concept of matter and the claim of materialism to explain the world around us.

Let us stay in physics for a moment and observe that, in addition to matter, physics defines another fundamental concept to account for the world around us, namely that of energy. The concept of energy is, in a way, the indispensable complement to the concept of matter: matter is that which is limited in space and has mass, i.e., that which is essentially inert and static; in contrast, energy is the capacity to modify a state of matter or to produce work leading to a movement of matter, i.e., that which allows matter to be in motion—without which nothing would ever happen. Let us note in passing that the concepts of matter and energy form a duality that has the same characteristics as the object-subject and matter-spirit dualities: in all these dualities, the first term is inert and passive, while the second is active. It is impossible not to further compare them to the duality between woman and man, which Western civilization has, for thousands of years, presented as an opposition between the passive feminine side and the active masculine side of the human species. Clearly, the relationship between the sexes has haunted Western civilization since its beginnings and is deeply rooted in its way of thinking.

For our present purposes, the crucial point to emphasize is that, of these two concepts, matter and energy, neither can exist without the other—at least in classical physics or in the world accessible to us through our senses: matter cannot exist without energy, without which nothing would happen and therefore nothing would exist; energy cannot exist without matter, without which nothing would be in motion. Just as the subject-object duality reflected an arbitrary distinction between entities that constantly interact with one another, the matter-energy duality is an arbitrary representation of the complex "substance" of the world, which inextricably mixes the properties of both what we are used to calling "matter" and what we are used to calling "energy." Albert Einstein had the genius to provide an explicit relationship between the two aspects of this "substance" of the world through his famous equation $E = mc^2$. Particle and high-energy physics now places energy rather than matter at the center of its description of the world; it routinely uses a measure of energy, the electron-volt, as a measure of the mass of elementary particles. Thus, without even going into the mysteries of quantum physics, which calls for even more profound changes in our way of thinking, a superficial examination of the concepts of modern physics is sufficient to demonstrate that the famous "inert matter," pure spatial expanse with no motion of its own that constitutes the foundation of Descartes's philosophy and of all modern thought, is only an arbitrary and fragmentary representation of reality by the human mind, a collective fiction that we continue to believe in for lack of anything better, and almost out of habit. Consequently, the materialism to which modern science is supposed to lead us naturally rests on foundations that are much less solid than they appear at first sight.

What about spirit? In Western civilization, the concepts of spirit and mind are opposed to those of matter and body; they designate either the immaterial soul proper to humans in the religious traditions—a meaning generally associated with the term "spirit"—or the mental faculties of humans in the scientific tradition—a meaning generally associated with the term "mind." But, as we have already mentioned in the previous chapter, in "primitive" hunter-gatherer societies as well as in other spiritual traditions, "spirit" or "soul" designates something much broader, a principle or breath of life that animates the whole cosmos. It is obvious that these concepts, by definition vague, immaterial, and inaccessible to observation and experimentation, cannot satisfy scientists; this is why, in science, "spirit" tends to be reduced to "mind," that is, to individual humans' mental faculties, which are accessible to observation and experimentation. For the rest, "spirit" or "soul" in their broader sense are relegated to the realm of individual beliefs, just as religious beliefs are.

It is difficult, however, not to draw a parallel between the concept of energy used in physics and that of spirit used in philosophical and religious traditions. Both concepts are opposed to that of matter, albeit in slightly different ways, to designate what is missing in matter, namely movement, life, consciousness. Matter without movement, without life, without consciousness, does not exist; it is pure abstraction. If one wishes to retain the abstract concept of matter, one must accept its counterpart in the form of energy or spirit. As a scientist and an atheist, what bothered me for a long time about the concept of spirit was its elusiveness and its religious anchorage. Today, it seems to me that the concept of matter is equally elusive and rooted in the religious tradition. The "spirit" or "soul" of animists and spiritualists is merely the active, immaterial side of what has been reified as inert matter; it is the equivalent of physicists' energy. Like spirit, energy remains mysterious; yet everyone agrees that it is everywhere—even, according to quantum physics, where there is no matter, in the immense void that immerses and penetrates all matter. If we accept this evidence, materialism and spiritualism cease to have consistency and meaning. For instance, you do not have to be a materialist to be an atheist; and you do not have to be a spiritualist to recognize the movement that runs through, animates, and connects all things in this world. Indeed, it is only by embracing at once the material and immaterial sides of the world around us that we can be fully alive, that is, participate in the general movement of life without artificially restricting the exercise of our physical and mental faculties.

In order for our worldview to keep pace with the evolving knowledge of contemporary physics, Bohm (1983) stressed the need to go beyond the "explicate order" described by classical physics and consider the "implicate order" of the world, a much broader order driven by the overall movement of the universe. According to Bohm, the implicate order is the total order of the universe considered as a single, indivisible whole across space and time, and thus it is contained implicitly in every single region of space and time. Philosopher Alfred Whitehead (1920) said no different when he asserted that nature is a process, that it is just another name for the creative force of existence, whose presence must be sought in the whole movement of the universe, in the distant past, in the present, and in the unrealized future. But if the overall movement of the universe is one and indivisible, it is this movement that must be considered fundamental, not matter, energy, or spirit. Matter, energy, and spirit are only concepts that seek to represent the material and immaterial aspects of the same universal movement; it is therefore futile to try to oppose them or to isolate them as separate realities.

The same applies to the distinction between living and non-living, animate and inanimate. What we consider to be inert matter is, in fact, in perpetual motion, and this motion is an integral part of the overall movement of the universe. Life is just one particular form of this overall movement. Bohm (1983, 194) gives this simple example to illustrate this truth:

> As the plant is formed, maintained and dissolved by the exchange of matter and energy with its environment, at which point can we say that there is a sharp

distinction between what is alive and what is not? Clearly, a molecule of carbon dioxide that crosses a cell boundary into a leaf does not suddenly "come alive" nor does a molecule of oxygen suddenly "die" when it is released to the atmosphere. Rather, life itself has to be regarded as belonging in some sense to a totality, including plant and environment. It may indeed be said that life is enfolded in the totality and that, even when it is not manifest, it is somehow "implicit" in what we generally call a situation in which there is no life.

But what is this overall movement of the universe that Bohm speaks of, if not the "spirit" or "soul" of the world that the animists had in mind? Stripped of its mysterious trappings in this way, the "soul" of the world ceases to appear disturbing to the scientific mind; it even becomes potentially accessible to scientific knowledge.

The supporters of materialism will no doubt reply that materialism has never denied the energy and movement of matter—movement was even the basis of the dialectical materialism proposed by Marx and Engels in the nineteenth century. But this argument, which I would have put forward myself some time ago, no longer satisfies me today because it pulls us backward rather than forward. Why then cling to matter as the foundation of the world, when physics itself has largely ceased to do so? Why present energy, motion, life, and consciousness as properties of matter, as, for example, Friedrich Engels (2012 [1925]) did in his famous book *Dialectics of Nature*, rather than the other way round? Atomistic materialism was essentially static; dialectical materialism made it dynamic; but today I see no longer any justification for the continued attachment to the abstract concept of matter shown by materialists. Materialists, it seems to me, cling to the primacy of matter more out of habit and opposition to religion than on the basis of sound philosophical and scientific arguments.

Unfortunately, modern society and science itself remain largely trapped in the straitjacket of the old duality between matter and spirit, and this duality clouds our minds to such an extent that there is great confusion about the nature of materialism and spiritualism. A great many people, including eminent scientists and philosophers, attribute properties to these ideologies that they do not have; as a result, they draw erroneous conclusions about the roots of the problems they face and the solutions to these problems, especially with regard to the current ecological crisis. One such misconception, which is extremely widespread today, is that the current ecological crisis is the result of the prevailing "materialism" of modern society, which aims to increase the production of material goods at the expense of nature. The corollary of this misconception is that, in order to overcome the ecological crisis, salvation would therefore lie in the development of spiritualist approaches that seek to free humans from material constraints. However, I will now show that modern "materialism" is only a disguised form of spiritualism. Spiritualism as such is therefore of no use in solving the problems created by modernity; some of the many spiritualist approaches may well accommodate, or even exacerbate them.

To understand this seemingly paradoxical claim, let us first remember that matter is an abstract concept created by Western civilization as one of the poles of a duality that opposes it to that of spirit or mind. In this duality, matter exists only in opposition to spirit or mind, just as spirit or mind exists only in opposition to matter. As the old wisecrack has it, "What is matter? Never mind! What is mind? No matter!" In this duality, matter is indispensable to spirit: without matter, spirit would have no concrete power since it would have no material on which to exert its action. Now we saw in Chapter 2 that the essence of modernity is precisely to assert the spiritual power of humans over inert matter, of which nature is made. Without matter stripped of spirit, this Promethean myth of modernity would collapse completely. In particular, the spirit-populated world of animism is incompatible with the unlimited exercise of human power over nature. Indeed, in a spirit-populated world, humans would have to deal with a host of other spirits all the time. They would have to find a delicate balance so as not to clash with them, and thus expose themselves to dangerous reprisals. The affirmation of the existence of inert matter, devoid of any spirit and animation, is therefore crucial to legitimize the historical mission entrusted to humans by Christianity and then modernity. Capitalism, as the economic expression of modernity, could never have conquered the world as it did without stripping it of the spirits that populated it in the minds of animists. Thus, a profound mystification takes place: modern materialism, which forcefully affirms the existence of inert matter, actually legitimizes the spiritual power of humans over this inert matter. Paradoxically, materialism is a glorification of the human mind or spiritual power. It is therefore a form of spiritualism, albeit a special one because it is masked by an appearance to the contrary.

The thesis I defend here may seem daring, and perhaps even shocking to some. However, it fits in perfectly with modern history. Is it by chance that the greatest materialists, including Marx, were also among the most ardent defenders of humans' transformative action on nature, which asserts the spiritual power of humans over the material world? But above all, my thesis is corroborated by the evidence of the deep religious roots of capitalism. Capitalism is traditionally seen as an economic system whose raison d'être is the production of material goods to increase the material well-being of humanity.

In reality, this is not the case. If there is one thing that Marx clearly demonstrated, it is that the primary purpose of capital is not to create use values, that is, material goods useful to humans, but to create exchange value, that is, abstract social wealth taking the universal form of money. Of course, exchange value must take the concrete form of use values, i.e., of useful material goods, exchanged on the market in order to be realized in its universal form of money, but this is a constraint for capital within which it is forced to move, not a goal. Capital has no other goal than to increase itself—it is, in a way, the economic manifestation of the Promethean will to power of modernity. What increases in this process of self-development is not so much the people and material objects necessary for its unfolding as abstract immaterial wealth in the form of money. Capital is a collective fiction created by humans, but a fiction so powerful that it governs their actions without most of them realizing why they act the way they do.

It could be argued that, far from being a collective fiction, capital is rather a tool in the hands of the wealthy who own it and whose purpose is to increase their personal power. There is no doubt that the owners and managers of capital benefit enormously in terms of wealth and social power. In particular, they have had a marked propensity recently to appropriate a growing share of social wealth for themselves, leading to a massive increase in economic inequalities over the past decades (Piketty 2013). Nevertheless, it would be a very serious mistake to reduce capital to a mere issue of unequal distribution of social wealth. As a matter of fact, the most ambitious capitalists reinvest most of their wealth in the production of new wealth, so that their personal share is only a tiny fraction of the profit of the capital they own. This was the case for most capitalists during the rise of industrial production in the eighteenth and nineteenth centuries, and it is still the case today for newcomers who create a new market or a new share of the market, like the new internet giants (Google, Apple, Facebook, Amazon, Microsoft, and the like) in their early days.

In his book *The Protestant Ethics and the Spirit of Capitalism*, early twentieth-century sociologist Max Weber (2011 [1904-1905]) demonstrated how capitalism's materialistic obsession with the production of material goods actually hid an abstract ascetic rationality inherited from Protestantism and directed toward increasing the glory of God. His study of the diverse currents of religious asceticism led him to this conclusion:

> To recapitulate, decisive again and again for our investigation was the conception of the religious "state of grace." Reappearing in all the denominations as a particular status, this state of grace separated people from the depravity of physical desires and from "this-world." (. . .) The possession of the state of grace (. . .) could

be acquired only through a *testifying to belief*. Sincere belief became apparent in specifically formed conduct unmistakably different from the style of life of the "natural" human being. There followed, for the person testifying to belief, a thrust to methodically supervise his or her state of grace. An organizing and directing of life ensued and, in the process, its manifold penetration by *asceticism*. As we noted, this ascetic style of life implied a *rational* formation of the entire being and the complete orientation of this being toward God's Will. (. . .) At its beginning, Christian asceticism had fled from the world into the realm of solitude. (. . .) Nonetheless, in retreating to the cloister, asceticism left the course of daily life in the world by and large in its natural and untamed state. But now Christian asceticism slammed the gates of the cloister, entered into the hustle and bustle of life, and undertook a new task: to saturate mundane, *everyday* life with its methodicalness. In the process, it sought to reorganize practical life into a rational life *in* the world rather than, as earlier, in the monastery. Yet this rational life in the world was *not of* this world or *for* this world. (Weber 2011, 156-157)

According to Weber, religious asceticism led directly to what he calls the "spirit of capitalism."

The "saint's everlasting rest" comes in the next world. On earth, in this life, in order to become certain of one's state of grace, a person must "work the works of Him who sent him, while it is day" (John 9:4). According to the will of God, which has been clearly revealed, *only activity*, not idleness and enjoyment, serves to increase His glory. (Weber 2011, 160)

   On the one hand, this-worldly Protestant asceticism fought with fury against the spontaneous *enjoyment* of possessions and constricted *consumption*, especially of luxury goods. On the other hand, it had the psychological effect of *freeing the acquisition of goods* from the constraints of the traditional economic ethic. In the process, ascetic Protestantism shattered the bonds restricting all striving for gain—not only by legalizing profit but also by perceiving it as desired by God (in the manner portrayed here). The struggle against the desires of the flesh and the attachment to external goods was *not*, as the Puritans explicitly attest (. . .), a struggle against rational acquisition; rather, Puritans challenged the irrational use of possessions. (Weber 2011, 169)

Thus, the original goal of increasing the glory of God gradually transformed itself into a new goal of accumulating goods, money, and capital. This makes sense since, after all, money and capital are only ever abstract manifestations of human labor dedicated to the glory of God. The accumulation of capital was not only permitted by Christianity; it became, with ascetic Protestantism, a *moral obligation* that follows directly from the mission assigned to humans by God. Max Weber's historical analysis shows, in a masterful way, how capitalism, in essence, is radically different from any form of material enjoyment of life; on the contrary, it responds to an ascetic will to work for the glory of the divine spirit in this world.

Writer Lewis Mumford (2010 [1934]) traces the origins of capitalism and mechanization even further back in time, to the monasteries of the Middle Ages, with similar conclusions:

> Within the walls of the monastery was sanctuary: under the rule of the order surprise and doubt and caprice and irregularity were put at bay. (...) The monasteries (...) helped to give human enterprise the regular collective beat and rhythm of the machine; for the clock is not merely a means of keeping track of the hours, but of synchronizing the actions of men. (...) Time-keeping passed into time-serving and time-accounting and time-rationing. As this took place, Eternity ceased gradually to serve as the measure and focus of human actions. The clock, not the steam-engine, is the key-machine of the modern industrial age. (...) The clock, moreover, is a piece of power-machinery whose "product" is seconds and minutes: by its essential nature it dissociated time from humane events and helped create the belief in an independent world of mathematically measured sequences: the special world of science. (Mumford 2010, 13-15)
>
> In still another way did the institutions of the Church perhaps prepare the way for the machine: in their contempt for the body. (...) Hating the body, the orthodox minds of the Middle Ages were prepared to do it violence. Instead of resenting the machines that could counterfeit this or that action of the body, they could welcome them. The forms of the machine were no more ugly or repulsive than the bodies of crippled and battered men and women, or, if they were repulsive and ugly, they were that much further away from being a temptation to the flesh. (Mumford 2010, 35-36)

Thus, far from the classical narrative that presents modernity as a break with medieval Christianity, modernity and capitalism are fundamentally an extension of Christianity in new forms. Modernity and capitalism have made it possible to realize on Earth the historical mission assigned to humans by the Judeo-Christian God, namely the domination of human spiritual power over natural matter. The accumulation of goods and money, which is the hallmark of capitalism, was originally conceived as an action dedicated to the glory of God, that is, as a work of Christian spirituality. Even when this religious justification disappears completely, the very nature of capital ensures the continuity of this particular form of spirituality that it conveys, unbeknown to its agents. In contemporary capitalism, everyone thinks they are contributing to the production and consumption of material goods useful to humankind, but few people are aware that behind this apparent objective lies the true meaning of this social enterprise: the accumulation of abstract social wealth in the form of money as a manifestation of the spiritual power of humans over nature. Capitalism has stripped the world of its soul, spirits, and gods, thereby contributing to promoting a materialistic worldview, but it has done so to tear us away from the matter that is supposed to constitute nature. Matter is a fiction, and capital is another fiction whose purpose is to assert the human spiritual power over the first—and, it goes without saying, especially the spiritual power of those humans who own capital.

It should now be clear why the solution to the ills of modernity and capitalism does not lie in spirituality per se: modernity and capitalism are already entirely immersed in spirituality and abstraction in their own way. The materialism they convey is a way of legitimizing the power of spirit and abstraction. But the "spirit" and "matter" of which capital is made are far removed from what we can experience as sentient human beings: they are essentially the spirit and matter that Descartes defined nearly four centuries ago, and which continue to govern our existence without our even realizing it.

What we need today to get out of this straitjacket is neither more spiritualism nor more materialism; what we need is to get rid of the hold that all of these collective fictions have on our lives and to reconnect with the sensible world within and around us, at once in all its materiality and in all its spirituality.

5

The underside of economic rationality and progress

"It's the economy, stupid!" This little phrase used by former US President Bill Clinton during his 1992 presidential campaign has become a cliché to remind us that, in the final analysis, it is the economy that governs our lives and that fine words are only listened to if they result in an improvement in our economic welfare. All the philosophical considerations in the world about the concepts of subject and object or spirit and matter will remain ineffective if they do not lead to a transformation of the economy on which our daily lives are based. I have already sketched out some of the characteristics of the modern social and economic order in which we live in the preceding chapters, but here I would like to take a closer look at some of the core beliefs that motivate our daily renewed confidence in this order and the consequences they have for modern society's relationship with the rest of nature.

Let us recall, first of all, that despite its apparent inevitability, capitalism—the social and economic order to which the vast majority of the world's human population is subjected today—is a product of human history, created by humans and reproduced, day after day, by their thoughts and actions. Nothing in this order is "natural" in the traditional sense that it is determined by immutable laws that preexist it. In fact, this order would have seemed completely delusional to our hunter-gatherer ancestors, and even to our more immediate ancestors. The social and economic order that governs our lives is entirely built on collective fictions. But these fictions are so ancient and powerful that they shape our subjective desires and consciousnesses and are embedded in the material world itself.

Harari (2011, 31-36) explained this very well:

People easily acknowledge that "primitive tribes" cement their social order by believing in ghosts and spirits, and gathering each full moon to dance together around the campfire. What we fail to appreciate is that our modern institutions function on exactly the same basis. Take for example the world of business corporations. Modern business-people and lawyers are, in fact, powerful sorcerers.

Nature That Makes Us Human. Michel Loreau, Oxford University Press. © Oxford University Press 2023.
DOI: 10.1093/oso/9780197628430.003.0006

The principal difference between them and tribal shamans is that modern lawyers tell far stranger tales. The legend of Peugeot affords us a good example. (. . .) In what sense can we say that Peugeot SA (the company's official name) exists? There are many Peugeot vehicles, but these are obviously not the company. Even if every Peugeot in the world were simultaneously junked and sold for scrap metal, Peugeot SA would not disappear. (. . .) The company owns factories, machinery and showrooms, and employs mechanics, accountants and secretaries, but all these together do not comprise Peugeot. A disaster might kill every single one of Peugeot's employees, and go on to destroy all of its assembly lines and executive offices. Even then, the company could borrow money, hire new employees, build new factories and buy new machinery. (. . .) In short, Peugeot SA seems to have no essential connection to the physical world. Does it really exist? Peugeot is a figment of our collective imagination. Lawyers call this a "legal fiction." It can't be pointed at; it is not a physical object. But it exists as a legal entity. (. . .) How exactly did Armand Peugeot, the man, create Peugeot, the company? In much the same way that priests and sorcerers have created gods and demons throughout history, and in which thousands of French curés were still creating Christ's body every Sunday in the parish churches. It all revolved around telling stories, and convincing people to believe them. (. . .) Ever since the Cognitive Revolution, Sapiens have been living in a dual reality. On the one hand, the objective reality of rivers, trees and lions; and on the other hand, the imagined reality of gods, nations and corporations. As time went by, the imagined reality became ever more powerful, so that today the very survival of rivers, trees and lions depends on the grace of imagined entities such as the United States and Google.

Money, which plays a pivotal role in modern economy, is also the product of an old collective fiction of humankind. Initially, when it was not yet serving as capital, money was currency, a universal means of social exchange between all sorts of commodities. Sociologist Michel Freitag (2008), however, reminds us that currency was more than a mere economic tool intended to achieve the commensurability of exchange values between various goods of a heterogeneous nature. Money as currency is also a political creation, invested with religious value.

While all exchange originally creates a specific link between the partners involved, a relationship of reciprocal dependence that takes the form of a debt and obligation not only between them, but also in the eyes of the whole community, the first meaning invested in the monetary symbol was that of an abolition or emancipation. But this abolition of the social and personal debt linked to any transaction did not simply take place in a magical or technical way, it implied the displacement or projection of the relationship of dependence into a new common relationship of subjection of the contracting parties to the political power, which henceforth assumed or guaranteed the liberating value of monetary payment. And thus, it was the effigy of this power, which ultimately exercised the liberating power, that was struck on one side of the coin, while the other side bore the quantified inscription of the liberating value of the coin, a value that was precisely guaranteed by the power. This is why we may say that money is a political creation, whose nature remains political. Money only frees subjects from their debt and their relationship of mutual dependence in the domain of their private *dominium* (. . .) by placing them in the field of a common public *imperium*. (Freitag 2008, 38)

The historical process of money creation also highlights another important point: while the social and economic order is made up of a set of collective fictions that all members of a society must share for it to function effectively, these collective fictions are ultimately based on power relations. A set of collective fictions only becomes a social order governing the lives of all the members of a society if it is legitimized and assumed by the political power. Money creation provides a good illustration of the thesis I put forward in Chapter 3 on the constitution of the modern subject, which achieves a new synthesis of the political subject, subjected to its sovereign, and the economic subject, bearer of individual action. By using money, the modern economic subject at once affirms the freedom of the Cartesian thinking subject, who can freely exchange goods on the market thanks to money, and his/her subjection to the dominant ideological, political, and economic order, which guarantees his/her freedom of action within the limits circumscribed by the authority of the state.

The social order created by modernity is particularly powerful and effective because it unites the ideological, political, and economic spheres into a coherent and homogeneous worldview, at the center of which the individual subject is enthroned. The individual is supposed to think freely and rationally (ideology), to choose freely and rationally those who govern him/her (politics), and to exchange freely and rationally on the market (economy). Rationality plays a particularly important role in modern economics. Contemporary economists have even created a fictional human species, ironically called *Homo economicus*, whose members behave as rational beings who maximize their utility (that is, the satisfaction they derive from the personal consumption of the goods and services they have purchased) in the market exchange. Economic theory shows that the free market is then the optimal, and therefore rational, way to ensure the distribution of goods among *Homo economicus*. Thus, everything nicely fits together: the market justifies the fiction of *Homo economicus* as a rational economic agent, and this fiction in turn justifies the market as a rational means of exchange.

The "reason" referred to in economic theory is instrumental reason, a specific form of reason inherited from market exchange and mathematical calculation. The term "reason" itself comes from the Latin word *ratio*, which means measurement, calculation, ratio. "Rational" in this sense is that which makes it possible to establish a quantitative relationship between different things. This is exactly what exchange value does in commodity exchange: exchange value establishes a quantitative relationship between goods of different qualities, which allows them to be compared with each other by erasing their intrinsic qualities. Exchange value, which takes the universal form of money, thus reduces all goods to an inert object that has lost its "soul" or quality and has only a quantitative scope. One cannot help but be struck by the similarity between the "matter" defined by Descartes, pure expanse in physical space, and the exchange value of commodity exchange, pure expanse in the abstract space of social wealth. The former is the reflection, in the world of philosophy, of the latter. The paradox is that matter, by definition, designates what is supposed to be purely material, while exchange value designates something eminently immaterial since its specific purpose is to free commodities from their material qualities. The dualistic concepts of modernity constantly present confusions of this type. The whole of reality seems inverted because the modern worldview gives preeminence to the thinking subject in the definition of what his thought is about, i.e., the material world of objects. The "object" is a fiction created by the one who arrogates to himself the position of "subject." Therefore, as we saw in the previous chapter, even "matter" is an abstract, immaterial concept that has little to do with the real substance of the world.

The essence of instrumental reason is to erase quality in favor of quantity, the concrete in favor of the abstract. Thus, it is economically "rational" to increase exchange value, which is the abstract quantity that erases quality in economic relations between people. Since capital is exchange value that increases itself, capital is eminently "rational" in the modern instrumental sense of the word. Note, however, that this form of rationality is largely tautological; it says nothing about the benefits that capital does or does not bring to people in flesh and blood. This is why it is important to bear in mind that the term "reason" that modernity uses and abuses covers at least two very different concepts, which economist and philosopher Serge Latouche (2004, 115) distinguishes as follows: "the rationality with which modernity wants to build the world is neither the wisdom of the Ancients nor the traditional and discursive forms of reason which, for my part, I gather under the concept of 'reasonable.' The rational concerns exclusively the calculating reason, for the domain of quantifiable objects. The reasonable concerns deliberative reason for the domain of non-quantifiable objects, and in particular ethical values."

Modernity, instrumental rationality, and capital go hand in hand with progress. Progress is, and remains, one of the great myths of modern society, a myth constantly invoked to legitimize the modern economic, political, and ideological order. There is no doubt that, in many respects, human life is more pleasant and comfortable today than it was a few centuries ago, at least in the world's richest countries. Despite the countless forms of violence that still exist, deaths by violent means have fallen considerably in recent centuries; healthcare is better and life expectancy longer; famines have practically disappeared in rich countries (though not in poor countries); and people enjoy greater freedom of action, which enables them to develop their individual capacities better. These real improvements, however, cannot be used as an excuse for the serious dysfunctions that modern society has created at the same time, whether in the form of increasingly extreme inequalities, both within and among countries, or the ecological crisis that we are entering head-on. The most fundamental problem with the notion of progress is that it is inherently limitless, that once trusted, it inevitably leads us into a never-ending headlong rush. Indeed, *"the belief in progress is self-fulfilling.* If we are convinced that the accumulation of knowledge, the improvement of techniques, the development of productive forces, the increase in the mastery of nature are *good things*, we act so that knowledge is transmitted and accumulated, so that the effects can be compared and increased. We give ourselves scales where indefinite growth becomes possible and relevant. This necessarily presupposes the *conviction* that the 'march forward' is an *amelioration* (from *melior*, better), and thus that it is a good thing" (Latouche 2004, 162).

Exactly the same is true of capital. Capital is a collective human enterprise based on shared confidence in an imaginary future. It cannot live without growth because capital is the process by which exchange value increases itself. But the growth of exchange value hinges on the collective belief that it will be realized, because the holders of capital only invest it if they believe that their investment will bring them a profit, and thus that the quantity of money they hold will increase. Thus, like progress, the growth on which capital is based is self-fulfilling. Credit plays an essential role in this process. The term "credit" speaks for itself as it comes from the Latin verb *credere*, which means to believe. Credit literally means belief. It is the lender's belief that the borrower will be able to repay his debt. But it is also the borrower's belief that he will be able to repay his debt in the future, and therefore the belief that he will get more wealth tomorrow than he borrows today. Thus, credit is based on the shared belief of the lender and the borrower that the future will be better than today. In turn, it becomes the means of realizing this belief since it offers the borrower the possibility of contributing to the growth of social wealth. Thus, credit is a privileged means by which the self-fulfilling prophecy of growth is realized.

The myth of progress creates the ideological conditions for the whole process, as it sustains the belief in a better future that mobilizes capital and credit. The history of capitalism is largely based on the myth of progress. Before the scientific revolution,

> because credit was limited, people had trouble financing new businesses. Because there were few new businesses, the economy did not grow. Because it did not grow, people assumed it never would, and those who had capital were wary of extending credit. The expectation of stagnation fulfilled itself. Then came the Scientific Revolution and the idea of progress. (. . .) Whoever believes in progress believes that geographical discoveries, technological inventions and organisational developments can increase the sum total of human production, trade and wealth. (. . .) Over the last 500 years, the idea of progress convinced people to put more and more trust in the future. This trust created credit; credit brought real economic growth; and growth strengthened the trust in the future and opened the way for even more credit. (Harari 2011, 346-347)

As long as this never-ending spiral of trust in the future, credit, and economic growth is accompanied by concrete positive impacts on human living conditions and the satisfaction of human needs, it appears as a virtuous cycle. But when it ceases to do so and comes up against the ecological limits of nature, as is the case today, its irrationality (in the sense of "unreasonable") tends to come to the fore. For, in the final analysis, it is the hope of a better life and a greater satisfaction of their needs that pushes people to accept and internalize the collective fictions of modernity.

One of the facets of the myth of progress, which largely contributed to the gradual acceptance of capital by the working class in the nineteenth century, is the belief that the enormous suffering endured by the working class during the first phase of capital's development was only temporary, and that the new capitalist economic order is a necessary historical step to create an affluent society in which everyone can flourish. This belief was promoted, not only by the rising bourgeoisie to establish its economic and political dominance, but also by its fiercest opponent, Marxism. The bourgeoisie believed that the historical stage of capitalism would continue indefinitely and that the promised abundance would be realized within it, while Marxism asserted that capitalism was only a transitional stage, which would lead to the real affluent society, communism. One cannot fail to be struck by the parallel between this promise, made by both capitalism and Marxism, of a future ideal world that will be realized on Earth after a life of hard work and suffering, and the promise made by Christianity of an ideal world that will be reached in heaven after a life of hard work and suffering. This promise is a continuation of the Promethean vision of humankind freeing itself from nature and the material constraints that run through Christianity, capitalism, and Marxism.

This promise is also a myth. Indeed, although it is particularly tenacious, this widespread belief does not stand up to rigorous scrutiny. One of the first economists who sought to make economic theory coherent in the early nineteenth century was Thomas Malthus. Synthesizing the ideas of his time, Malthus postulated that the goal of economics was the creation of social wealth, and he defined wealth as "the material objects which are necessary, useful or agreeable to man, and which require him to make some effort to produce or appropriate them" (cited in Naredo 2003, 117). It can be seen from this definition that the very concept of wealth excludes the products of nature and thus implicitly presupposes the economic notions of utility, value, labor, and production, even though these were supposed to be justified by the creation of wealth. This conceptual tautology has far more important consequences than may appear at first sight. As economist José Manuel Naredo (2003, 119) rightly pointed out, "if the object of economic science is *wealth* and not useful things in general, and if it aims to increase the subset of useful things that are, by definition, *scarce* and not all useful things, this objective will lead to *scarcity* and not to *abundance*. No matter how much *wealth* is increased, it will still be what it is by definition—*scarce* and *laborious* to obtain."

The dominant discourse is that capitalism is a means to an affluent society. The abstract social wealth that capital increases, however, is intrinsically based on scarcity; consequently, capital can only ever create scarcity. This leads to the paradox, already noted by Sahlins (2017 [1972]), that the archaic hunter-gatherer societies were, in fact, closer to affluence than modern capitalist society. Much more: the creation of modern social wealth presupposes the destruction of archaic affluence.

If *production* is defined as "the creation of objects that constitute wealth," it is easy to realise that the most effective way to boost one's production is to transform into wealth useful things that were not wealth before, by making them *scarce* or by demanding *efforts* that were not necessary before to obtain them. Thus, destroying the original context in which useful things were obtained abundantly and freely is sufficient to inflate the sphere of the *production and consumption of wealth*. This destruction has taken place and continues to take place, either by privatising and monopolising useful things that were originally abundant or that are sources of renewable resources, or by provoking their scarcity through an increased demand and a gradual depletion of their renewal capacity, or by shifting tastes and needs from the use of the abundant and renewable to that of the limited and scarce. Most of the time, in fact, these mechanisms overlap and act in concert. (Naredo 2003, 121-122)

It is undeniable that the quantity of material goods produced by capital has increased considerably throughout history. But, on the one hand, the human population has also increased and, on the other hand, the use value of these goods is only distantly related to the actual needs of the people who consume them. Many "goods" are in fact "evils" because they serve to compensate for frustrated fundamental human needs. They create a spiral of dependence of people on things whose ultimate purpose is not to satisfy their fundamental needs, but instead to produce abstract social wealth in the form of money. The best way to increase abstract social wealth is precisely to fail to satisfy fundamental human needs and to produce substitutes that temporarily compensate for the frustration caused by this lack of satisfaction, but that maintain it in the long term—this is the principle of drugs. Thus, in modern society, a paradoxical situation is created in which an excess of material goods is accompanied by a growing dissatisfaction of the fundamental needs of a significant part of the population.

The vocation of capital is to produce and reproduce scarcity because scarcity is the basis of exchange value: without scarcity of a good, there is no need for labor to produce it, and therefore no exchange value since everyone can obtain it at will. What matters for capital is not so much scarcity per se as the labor required to produce commodities, which essentially gives the measure of their exchange value. As long as people can get what they need without labor—as is the case in hunter-gatherer societies—both capital and labor appear to be pure folly. Thus,

the Hadza, tutored by life and not by anthropology, reject the neolithic revolution in order to keep their leisure. Although surrounded by cultivators, they have until recently refused to take up agriculture themselves, "mainly on the grounds that this would involve too much hard work." In this they are like the Bushmen, who respond to the neolithic question with another: "Why would we plant, when there are so many mongomongo nuts in the world?" (. . .) Extrapolating from ethnography to prehistory, one may say as much for the neolithic as John Stuart Mill said of all labor-saving devices, that never was one invented that saved anyone a minute's labor. The neolithic saw no particular improvement over the paleolithic in the amount of time required per capita for the production of subsistence; probably, with the advent of agriculture, people had to work harder. (Sahlins 2017 [1972], 26, 34)

The hunter-gatherers' aversion to work partly explains why the colonial expansion of capitalism into territories occupied by hunter-gatherer peoples was often carried out in bloodshed: it was necessary first to extirpate any form of original abundance and to create the conditions of scarcity that forced people to depend on their labor for survival.

We have already noted that the scarcity on which capital is based can nevertheless be accompanied by an excess of commodities produced, and thus an apparent material abundance. To understand this paradox, we need to understand that the scarcity that forms the basis of exchange value is *relative*—relative to other goods and to the wants or "preferences" that people have for different goods. Modern economic theory assumes that all goods are substitutable since they can all be reduced to a single measure: exchange value. It also assumes that economic agents have unlimited wants, but that they nevertheless have "preferences" that allow the relative desirability of different goods to be compared. In this theory, therefore, there can be no question of either natural, objective human needs or natural, objective scarcity. The *Homo economicus* of modern economics is the transposition of Descartes's thinking subject to the economic sphere, that is, the human individual who frees him- or herself from the material world in order to assert his or her free will and who can therefore in no way be limited by this material world. Exchange value is the economic expression of this "liberation" of humans from nature because it erases all the natural, objective qualities of commodities and reduces the latter to an abstract social wealth entirely created by humans. Exchange value therefore erases all natural, objective human needs and recognizes only "preferences," that is, subjective choices that are free of any normative constraints and perfectly substitutable with each other: "What creates *value* and thus the unified economic world of value is the erasure of objective *need* in the principial establishment of a double substitutability of *wants* among themselves in subjective choices freed from all normative constraints, and of goods among themselves in their generalised exchangeability, which removes them from their concrete normative assignments." Thus, "the economy does not begin where there is objective scarcity, but it creates (relative) scarcity wherever it imposes itself as a form of production and exchange, establishing by postulate an unlimited virtual mobility of all goods. From this it can also be said that economic freedom is primarily the freedom of goods, rather than of people, and that this freedom begins where goods are 'liberated' or emancipated from all the normative social constraints that assigned them a place in social life" (Freitag 2008, 79).

Just like the scarcity that is the basis of exchange value, the social wealth that results from it is relative. In modern society, the satisfaction of wants depends not only on the utility of the goods purchased, but also and above all on the social prestige they enjoy. It is not so much the absolute income of an individual that determines his or her wealth as his or her relative income, that is, his or her place in the social hierarchy of wealth. As nineteenth-century writer John Ruskin (1967 [1860], 30) observed, "The force of the guinea you have in your pocket depends wholly on the default of a guinea in your neighbour's pocket. If he did not want it, it would be of no use to you; the degree of power it possesses depends accurately upon the need or desire he has for it, — and the art of making yourself rich, in the ordinary mercantile economist's sense, is therefore equally and necessarily the art of keeping your neighbour poor." Thus, the wealth on which modern society is based necessarily goes hand in hand with scarcity, shortage, and poverty. Social inequalities are not a sign of temporary dysfunction, they are inscribed in the very foundations of capitalism, just as they were in the slave and feudal societies that preceded it. The domination of nature by humans is inextricably linked to the domination of humans by other humans. Both forms of domination are equally based on the denial of the human body and natural needs and the exaltation of human spiritual power.

The relative nature of scarcity and social wealth has many other deleterious consequences. Human wants stimulated by capital are by nature insatiable: no matter how rich a person is, he or she will often aspire to become even richer, either to climb to the top of the social ladder of wealth or to stay there. This well-known modern pathology is all the more developed the higher the position that person occupies in the wealth hierarchy, making the accumulation of wealth an unlimited process. As economist Herman Daly (1977, 41) pointed out,

> the implication of the doctrines of the relativity of scarcity and the insatiability of wants is growthmania. If there is no absolute scarcity to limit the possibility of growth (we can always substitute relatively abundant resources for relatively scarce ones), and no merely relative or trivial wants to limit the desirability of growth (wants in general are infinite and all wants are worthy of and capable of satisfaction by aggregate growth, even if based solely on invidious comparison), the "growth forever and the more the better" is the logical consequence.

Another consequence of the relative nature of scarcity and social wealth is that the internal mechanisms of the market are unable, by themselves, to solve the ecological problems they generate. One of the main principles used in natural resource and environmental economics today is that of "internalization of externalities." This barbaric term means that the damage caused indirectly to third parties through either an increased scarcity of natural resources or environmental degradation (the so-called externalities) should be "internalized," that is, included in the selling price of the goods whose production is responsible for these adverse effects. But the market cannot, by itself, achieve this internalization because it is based on relative scarcity, not absolute scarcity. Although the "internalization of externalities" may be a technical means in the implementation of public policies aimed at remedying the absolute scarcity of certain natural resources, this approach obscures the most essential initial step in the process for it to work, namely externalizing the costs outside the market:

> The market cannot, by itself, keep aggregate throughput below ecological limits, conserve resources for future generations, avoid gross inequities in wealth and income distribution, or prevent overpopulation. (. . .) Instead of internalizing external costs, the idea is to externalize them, that is, to take from the market sphere the possibility of incurring costs that it is unable to perceive or evaluate. Benefits and costs that do not register themselves as conscious short-run pleasure or pain at an individual level but that are organic, with interdependencies far exceeding market relationships, must be dealt with outside the market and must result in constraints on the market. (Daly 1977, 89)

It is only in a second stage, once these constraints have been decided upon as a result of stakeholder consultation or the definition of a public policy independent of the market, that these constraints can possibly be implemented through market mechanisms.

Capital has liberated human beings from the old fetters that bound them to their local community, family, or plot of land, and transformed them into "free citizens," masters of their personal destiny. But these free citizens are, at the same time, bound hand and foot to an extraordinarily powerful set of collective fictions which distracts them from the satisfaction of their natural needs, which tends to reduce the entirety of their social life and relations to nature to the production of a single abstract form of social wealth, and which leads them into a headlong rush toward the unlimited increase of this abstract form of social wealth. Only by becoming aware of their condition as freely consented slaves of these collective fictions can humans begin to free themselves from these fictions, find their true place in this nature they tried to escape or dominate, and build a true freedom, based on the satisfaction of their needs and the fulfillment of their human qualities on a living Earth.

6

Journey to the center of the modern world

All the core beliefs that underlie modernity—the subject-object duality, the spirit-matter duality, the myth of progress—bring us back to the same basic premise: humans, through their spiritual power, occupy a central place in the world. Humankind was created in the image of the Jewish or Christian God to dominate nature; everything in the earthly world we know is at its service. The idea that humans are at the center of the world—known as "anthropocentrism"—is both an obvious and a paradoxical product of modernity. Obvious, because the essence of modernity is precisely the affirmation of the privileged place and historical mission of humans in the world. Paradoxical because, at the same time, the scientific revolution brought about by modernity has constantly undermined this anthropocentric postulate. Astronomy first showed that the Earth was only one planet among others revolving around the sun, and not the other way around as had been believed until then; then biology showed that the human species was only a twig among many others in the teeming tree of life on Earth—what is more, a descendant of the ape, that much-vilified beast—and not the product of any divine will. But anthropocentrism is tenacious; it often lurks where we least expect it, including in the beliefs of many contemporary philosophers and scientists, however enlightened they may be in their fields of expertise.

Anthropocentrism is closely related to another product of modernity, namely humanism. The difference between the two doctrines is slight, but humanism enjoys far greater prestige today. The reason for this flattering reputation is probably to be found in the common association between humanism and the Renaissance. Yet Renaissance humanism had little to do with contemporary humanism. In the late Middle Ages and Renaissance, the term "humanist" was used to describe a man of letters who studied and taught the "humanities," that is, Latin and Greek grammar and rhetoric. It was only much later that humanism came to take on a philosophical meaning and to designate the doctrine that takes the human person as its end. Somewhat ironically, the first to give it this philosophical meaning seems to have been the anarchist

Nature That Makes Us Human. Michel Loreau, Oxford University Press. © Oxford University Press 2023.
DOI: 10.1093/oso/9780197628430.003.0007

Pierre-Joseph Proudhon (2002 [1846], 521) in the mid-nineteenth century, not to defend it, but on the contrary to reject it "as tending invincibly, through the deification of humanity, to a religious restoration." The *Merriam-Webster Dictionary* provides the following definition of the current philosophical meaning of humanism: "a doctrine, attitude, or way of life centered on human interests or values," and more especially, "a philosophy that usually rejects supernaturalism and stresses an individual's dignity and worth and capacity for self-realization through reason."

As this definition shows, there is good and bad in humanism. The self-realization of the human person should undoubtedly be one of the fundamental objectives of our social activity—we will return to this in Part II of this book. The first part of the definition, however, closely links humanism to anthropocentrism: if all our attention and action is focused on human interests and values, humans are in fact at the center of the world; the world revolves around them and is at their service. Since this part of the definition corresponds to the original meaning of the term, it is difficult not to agree with Proudhon's criticism of humanism, despite all the good intentions of those who claim to be humanists.

Humanism and anthropocentrism have been the ideological price to pay for the development of the human person that has taken place under capitalism. To be sure, capital has enslaved modern humans to powerful collective fictions, as we have seen at length in the preceding chapters. But by freeing the individual from their old tribal, communal, or familial attachments, it has also allowed them to experience their independence and creativity as a person beyond what was possible in earlier societies. At the same time, it has created a new form of interdependence between these individuals through the global market, thus laying the foundation for a global human community across personal or national beliefs and boundaries. These changes in the forms of human sociality are arguably the most significant advances that capital has made to humanity—more significant than the much-touted "material progress" that contributes to the destruction of ecosystems on which humanity depends and that may well be only ephemeral. By fleshing out the modern thinking subject in the economic sphere, capital has created a national, and then transnational, community of people free of their previous social attachments but highly interconnected. Recent developments in computing and telecommunications have accentuated this historical trend by making it subjectively manifest to the individuals themselves. The creation of the World Wide Web, the emergence of virtual social networks, and the advent of mobile phones have contributed powerfully to reinforcing both the sense of independence—and even isolation—of individuals on a small scale and that of their interdependence and belonging to the same human community on a large scale.

But the changes wrought by capital in the forms of human sociality have only served to increase the power of the human species and, within it, of those who hold economic power. They have been largely at the expense of the weaker segments of human society and other species, most of which have experienced an unprecedented deterioration in their living conditions and are now threatened with extinction. The avowed objective of modernity is the complete humanization of the world, the transformation of nature into a gigantic machine at the service of humans. There is also an implicit, undisclosed objective, that of putting this gigantic machine at the service of the accumulation of abstract social wealth and the power of those who possess it. Thus, the tremendous social development that has taken place under capitalism is extremely asymmetrical; it benefits exclusively the human species and, indirectly, the few species it has domesticated and, within the human species, primarily the holders of economic power. By emphasizing the "fulfillment of the human person" and the "respect for his dignity," humanism strikes a chord that resonates with every human being, but by taking "man as its end and supreme value," it legitimizes modernity's objective of increasing human power. Perhaps even worse, by combining the two aspects, it gives a veneer of respectability to this objective.

Humanism and anthropocentrism were heavily criticized by several currents of thought at the end of the past century. In particular, the animal liberation movement accused humanism of being a form of "speciesism" that arbitrarily discriminates against sentient beings of other animal species and thus should be condemned in the same way as racism and sexism (Singer 1975). Other currents challenged humanism more broadly, developing the idea that not only sentient animals, but also plants, species, ecosystems, and even the biosphere as a whole have "intrinsic value," that is, they should be considered as ends in themselves, irrespective of the utility they may represent for humans (Rolston 1988). These currents thus proposed to elevate the living world and its multiple components to the same status as humans in modern philosophy. This non-utilitarian approach forms the basis of environmental ethics, a discipline that seeks to define new ethical principles to govern our attitude toward the non-human living world. It also played an important role in the development of the nature conservation movements in the twentieth century, which focused on protecting species regardless of their use or monetary value (Adams 2004).

Environmental ethics, however, is not unanimously accepted in contemporary nature conservation movements. Since their inception, these movements have been riven by a conflict between utilitarian and non-utilitarian tendencies. In contrast to the ethical approach, the utilitarian approach considers the non-human living world as a set of resources, which should be managed prudently in order to avoid scarcity, as the latter could have harmful economic and social consequences for human societies. This utilitarian approach has its roots in natural resource and environmental economics. It is currently enjoying great popularity, linked to the recent craze for "ecosystem services." Human societies derive a large number of direct and indirect benefits from ecosystems; these benefits have been viewed as "services" provided by ecosystems to humans, by analogy with the notion of service used in economics (Millennium Ecosystem Assessment 2005). Although the new ecosystem service approach is broader than the classical approach of natural resource economics, it shares the same utilitarian and anthropocentric perspective, according to which biodiversity only matters as a means for humans.

Why has environmental ethics failed to offer a sufficiently robust and credible alternative to utilitarianism and the anthropocentrism that underlies it? First of all, ethics only remotely affects individual and collective decision-making processes, which are largely governed by short-term, human-centered interests in present-day society. But there are also other reasons for this, which have to do with the internal limitations of environmental ethics itself. Environmental ethics essentially proposes to extend the traditional boundaries of ethics to the non-human world, although the proposed new boundaries vary widely depending on the authors and the criteria used. One of the key concepts used to justify this extension is that of the intrinsic value of non-human entities. Modern classical ethics also uses this concept, but attributes it exclusively to humans. Philosopher Holmes Rolston (1988, 340) summarized the difference between traditional human-centered ethics and environmental ethics as follows: "Kant knew something about others, but, eminent ethicist though he was, the only others he could see were other humans, others who could say 'I.' Environmental ethics calls for seeing nonhumans, for seeing the biosphere, the Earth, ecosystem communities, fauna, flora, natural kinds that cannot say 'I' but in which there is formed integrity, objective value independent of subjective value." While most environmental ethicists share Rolston's view that non-human others—or at least some of them—should be recognized as having intrinsic value, the nature and source of this value have been hotly debated. In particular, there has been a confrontation between an objectivist viewpoint, which conceives of intrinsic value as an objective property of the object being evaluated, independent of the evaluator, and a subjectivist view, which sees intrinsic value as a subjective property of the human subject who evaluates.

However obscure it may seem to the uninformed reader, this debate is revealing of the limits of environmental ethics, which has failed to break out of the philosophical straitjacket of modernity, and in particular of its characteristic duality between humans and nature. In this debate, humans have been implicitly or explicitly placed outside nature, so that the problem represented by the objective or subjective nature of intrinsic value becomes critical and, indeed, insoluble. If humans are regarded as an integral part of nature, however, the problem ceases to be one. Ecology shows that humans, like any other living beings, are nodes in a complex network of interactions between the various components of the biosphere. The attribution of value by humans to other living beings is but one particular aspect of these interactions. The fact that humans, as valuing subjects, attribute value to other components of their environment, whether human or non-human, does not mean that they do so arbitrarily, independently of the objective properties of these components, nor does it mean that these components do not also behave as valuing subjects. Each species evaluates its environment in its own way, according to its cognitive capacity and ecology. As I have already noted in Chapter 3, the concepts of subject and object, insofar as they have any use, are nothing more than roles played, in a given relationship at a given moment, by interacting entities. These roles say nothing about the intrinsic nature or status of the entities in question.

Objectivist environmental ethicists have used the existence of "objective," human-independent value relationships between non-human entities as a decisive argument for the fact that the latter should be the object of moral consideration by humans. Yet this argument is clearly flawed. Indeed, what is perceived as "good" for one entity is not necessarily perceived as "good" for humans or other entities, as attested by the fact that antagonistic relationships are exceedingly common in nature, whether between prey and their predators, between hosts and their parasites, between hosts and their pathogens, or between species that compete for shared resources. Only values recognized or ascribed by humans can form the basis of environmental ethics since ethics is a human construct, which is designed to govern the social world of humans. Therefore, any attempt to eliminate the role of human subjectivity in the attitude that humans should adopt toward the rest of nature is doomed to failure.

Another limitation of environmental ethics is that, like classical ethics and modern philosophy, it relies heavily on rationality. For instance, environmental ethicist Paul Taylor (1981, 202-203) insists that moral commitment is moral only insofar as it is "a disinterested matter of principle. It is this feature that distinguishes the attitude of respect for nature from the set of feelings and dispositions that comprise the love of nature. The latter stems from one's personal interest in and response to the natural world. (. . .) Respect for nature is an attitude we believe all moral agents ought to have simply as moral agents, regardless of whether or not they also love nature. (. . .) To put it in a Kantian way, to adopt the attitude of respect for nature is to take a stance that one wills it to be a universal law for all rational beings." Of course, any moral consideration that aspires to universality must involve rationality, but to disconnect it completely from its emotional foundations is not only ineffective, but also illusory. As we saw in Chapter 1, recent work in the fields of neuroscience and human and animal psychology has clearly established that moral behavior is rooted in emotions and that it also exists, in primitive form, in non-human mammals. While rationality does allow useful generalizations beyond immediate emotional responses, it does not generate morality. Therefore, the emphasis of environmental ethics on pure rationality, on "disinterested matters of principle" and on the "objective" nature of intrinsic value, greatly reduces its capacity to serve as a guide for effective nature conservation. Reason and emotion cannot be separated, or else we condemn ourselves to schizophrenia or inaction.

Utilitarian approaches to nature conservation also invoke rationality, but in a more restricted sense. In contrast to ethics, which emphasizes "disinterested matters of principle," modern economics and utilitarianism invoke a purely instrumental form of rationality that serves the individual "interest" of human economic agents. It is no surprise that utilitarian approaches have found a favorable echo since modern economics is largely based on the maximization of economic utility, as recalled in Chapter 5. This explains why the new ecosystem service approach has been so popular during the past twenty years or so. Although this approach is broadly integrative (it even seeks to include the intrinsic values of biodiversity under the category of "cultural ecosystem services," while intrinsic values are supposed to be non-utilitarian forms of value), it is fundamentally utilitarian and anthropocentric, at least in concept, since it assumes that nature is at the service of humans. The common practice of economic valuation of ecosystem services makes it even utilitarian in the economic sense of the term as it incorporates ecosystems in the economy of human societies as a "natural capital" that provides humans with a flow of services with monetary value. The rationale for doing so is that giving ecosystem services a monetary value is expected to encourage economic actors to factor the benefits that human societies derive from ecosystems into the cost of goods and services that are traded on the market. If biodiversity and ecosystems have economic value, they are more likely to be preserved, either as a worthy source of income or as a means to reduce the costs inherent in environmental degradation.

But there are also considerable dangers in reducing nature to a capital that provides marketable services (Silvertown 2015). In particular, there will always be a great danger that short-term economic imperatives will override fundamental human needs and nature conservation. As biologist Edward Wilson (1992, 348) noted, ecosystem "services are important to human welfare. But they cannot form the whole foundation of an enduring environmental ethic. If a price can be put on something, that something can be devalued, sold, and discarded." One may even wonder if, by fostering the economic valuation of nature, the ecosystem service framework is not unwittingly preparing the ground for further infringements of capital on natural ecosystems. In comparison with these dangers and the high expectations placed in the ecosystem service approach, the economic valuation of ecosystem services has had surprisingly little impact on decision-making so far (Laurans et al. 2013), perhaps because this approach is still too coarse, or because economic valuation plays a more minor role in the decision-making process than assumed by economists.

The opposition between utilitarian and non-utilitarian approaches, between economics and ethics, between instrumental and intrinsic values, leads to an apparently insoluble philosophical dilemma: either human subjectivity is the source of all value, in which case value seems destined to be purely instrumental since the entities to which it is attributed have value only insofar as they are of interest to humans; or intrinsic values are objective properties of the entities under consideration, independent of human subjectivity, in which case they cannot suffice to justify the moral consideration that humans are supposed to give to these entities. In reality, this opposition and the dilemma it creates stem from the philosophical presuppositions of modernity, and in particular from the dualistic dichotomies between subject and object, and between spirit and matter. According to these dichotomies, what belongs to the operation of the human mind is subjective, and what belongs to the independent properties of matter is objective. Consequently, subjectivity is generally associated with the realm of the particular and the arbitrary, while objectivity defines the realm of the universal and the natural. This dualistic representation of the world ignores the fact that humans think and act as subjects on the basis of a universal biological human nature modulated by culture. Human subjectivity therefore necessarily has an objective, natural, universal dimension. Conversely, as we have seen in Chapter 3, the fact of being in the position of an object in relation to a subject does not imply the absence of particularity, spirituality, or freedom.

This opposition is also based on a dichotomy between the satisfaction of human needs—reduced to a lowly material aspect in the modern worldview—and moral or altruistic behavior—usually regarded, by contrast, as a noble product of the human mind. This dichotomy manifests itself in the widespread assumption that humans, like all living beings, are naturally selfish, so that the satisfaction of their needs necessarily implies treating others as instruments of this satisfaction, unless social and moral constraints forbid it. This assumption, however, results from a fundamental misinterpretation of evolutionary biology, which confuses the psychological motivation of living beings with the process of natural selection that affects the transmission of their genetic or cultural traits from one generation to another. The fact that cooperation, altruism, and morality can be favored by natural selection does not make them selfish behaviors. Charles Darwin (2011 [1871], 68, 78), the father of modern evolutionary theory, who is often wrongly attributed with a worldview based on selfish self-interest, was very clear on this issue: "But if, as appears to be the case, sympathy is strictly an instinct, its exercise would give direct pleasure, in the same manner as the exercise (. . .) of almost every other instinct. (. . .) Thus the reproach is removed of laying the foundation of the noblest part of our nature in the base principle of selfishness; unless, indeed, the satisfaction which every animal feels, when it follows its proper instincts, and the dissatisfaction felt when prevented, be called selfish."

As I mentioned in Chapter 1, recent work in neuroscience and human and animal psychology leaves no doubt that empathy (a term that has replaced the term "sympathy" used by Darwin and his contemporaries) is a fundamental characteristic of the human species, as well as other vertebrate species, that has evolved in response to their sociality. Empathy implies some form of identification with the other—whether or not that other belongs to the same species—and thus recognition of the other as an end in itself. In other words, empathy implies attributing intrinsic value to the other, whether consciously or unconsciously. We also know that empathy first involves sensory and emotional responses before any form of rational judgment, thus fully confirming Darwin's intuition about its instinctive nature. A recent experiment even suggests that humans are spontaneously empathetic and cooperative in the absence of rational judgment, and that selfishness is a consequence of the importance taken by rationality in adulthood (Rand et al. 2012). The remarkable conclusion that follows is that the intrinsic value we attribute to other beings, human or non-human, is embedded in our bodies. In other words, it is a "subjective" response that has an "objective" existence prior to any intellectual and rational processing. This shows again how artificial are the boundaries that are supposed to separate what belongs to "subject" and "object."

We will see in Part II of this book that humans do have a human nature, which can be defined by the totality of their biological inheritance and which includes in particular a finite set of fundamental human needs. One of the most striking features of these fundamental needs is that they extend far beyond the physiological or subsistence needs on which the emphasis has traditionally been placed. In particular, they include a wide range of needs related to relationships with others and self-fulfillment, the satisfaction of which relies entirely on non-utilitarian interactions with other living beings, whether human or non-human. Like empathy, the existence of fundamental human needs based on non-utilitarian interactions with other living beings destroys the separation between self-fulfillment and fulfillment of the other, since the fulfillment of the other is the condition for my own fulfillment, and vice versa. Thus, the satisfaction of fundamental human needs is fully compatible with the recognition of intrinsic values in the human and non-human world (Loreau 2014).

As a matter of fact, the distinction between instrumental and intrinsic values dissolves into a continuum without clear boundaries. At one extreme, things that are used or consumed to satisfy the basic needs of subsistence and protection clearly have an instrumental value, as they enter into a utilitarian relationship. At the other extreme, human and non-human persons who are respected, honored, and loved in order to satisfy the needs for affection and self-fulfillment are clearly endowed with intrinsic value, for it is their very existence as independent subjects that enables the satisfaction of these needs. But most, if not all, of the entities with which we interact are probably endowed with a double value, both instrumental and intrinsic. For instance, there is no good reason to consider that the plant and animal creatures we use for food are not worthy of respect, and therefore have no intrinsic value. Respect for hunted animals is widespread in hunter-gatherer societies. Although modern society places an almost pathological emphasis on utilitarian relationships, respect for non-human creatures is present in the depths of each of us, and only disappears from our consciousness as a result of a long process of education—in fact, erasure—during our childhood.

The new light that empathy and fundamental human needs shed on human nature helps to resolve the debate about anthropocentrism that has raged in environmental ethics. Environmental ethicists have rightly criticized classical ethics for its anthropocentrism, in that it places humans at the center of the world. By contrast, they have focused on developing—wrongly, in my view—new approaches that place non-human entities at the center of the world, whether in the form of "biocentrism" (Taylor 1981) or "ecocentrism" (Rolston 1988). But there is no more valid reason to place sentient animals or ecosystems than people at the center of the world. And why do we need to define a center of the world in the first place?

At first glance, it might seem that my proposal to focus on fundamental human needs takes us right back to classical anthropocentrism. But this is not the case. Any ethical or value system established by humans is inevitably based on the specific capacity of humans to perceive, understand, and evaluate the world around them. The fact that other creatures may do the same in their own manner does not affect this basic observation in any way. But the world that humans seek to perceive, understand, and evaluate is by no means limited to their social world alone; consequently, there is no need for it to be centered on them. Clearly, we can take responsibility for our own evaluation of the world around us without seeing ourselves as the center of this world. In other words, while the values underlying human ethics are necessarily "anthropogenic," that is, created by humans, they are not necessarily "anthropocentric," that is, centered on humans (Maris 2010).

The anthropocentric value system established by modern Western civilization results from the artificial division of the world into thinking human "subjects" and inert material "objects." Spirit versus matter, subject versus object, reason versus emotion, culture versus nature, intrinsic value versus instrumental value: modern Western civilization has an immoderate predilection for dualism. Unfortunately, these dualities are fictions that do not only exist in our minds; they also manifest themselves in the concrete reality of our action in the world. The global ecological crisis we have entered is the historical product of the separation between humanity and nature that results precisely from this dualism. Humans continue to destroy biodiversity and ecosystems to unprecedented degrees, largely because modernity's representation of the world teaches them that they are different from the rest of nature and that they have the mission to dominate it. To overcome the current ecological crisis, the first thing to do is to get rid of this outdated and deleterious worldview.

PART II
WHERE HUMANS AND
NATURE ARE ONE

7

Letting nature touch us

The modern program of human domination over nature is based on a set of powerful collective fictions that take us away from nature and, in doing so, lead us to destroy the natural foundations on which modern society itself has developed. Thus, building a new relationship between humans and nature requires some radical changes in the worldview we have inherited from modernity. In this and the following chapters, I will examine some of the most important shifts that need to take place in the way we experience and view the world and ourselves if the human species is to persist and flourish as an integral part of life on Earth.

The most obvious aspect that needs to change if we are to build a new relationship between humans and nature is the way we relate to "nature" as it is commonly defined, that is, to the living and non-living beings around us. We saw in Part I of this book that the modern worldview has de-animated the non-human world to glorify the active role and supremacy of humans by contrast. But the real world around us is animated, and well indeed; it is a world of constant flux and movement, of gigantic physical forces that have created the mountains and rivers as we know them and are rapidly changing the global climate, of a myriad of living beings that have evolved and transformed the Earth for billions of years before the first human appeared, and of sentient animals that, like us, possess language, thought, and empathy. Modernity's claim that humans are animated subjects in a world of de-animated objects is a particularly pernicious fiction that is completely out of step with the real world. If we are to establish a sustainable relationship with the rest of nature, we clearly need to abandon this antiquated fiction and embrace nature as a fully animated world.

Humans, however, are often so steeped in their beliefs—especially the oldest collective beliefs that have shaped social consciousness for hundreds, or even thousands of years—that intellectual clarity will not be enough to achieve this. The first thing to do is to simply stop believing and get back in touch with the world as it is—just breathe, walk, feel the breeze on your cheek, touch a tree, smell a flower, listen to birdsong, watch the stars shine in the sky or a bee pick a flower. I guess everyone has already experienced how inner tensions are relieved, an inner calm establishes itself, and the perception of the

Nature That Makes Us Human. Michel Loreau, Oxford University Press. © Oxford University Press 2023.
DOI: 10.1093/oso/9780197628430.003.0008

outside world becomes broader and clearer when being outdoors in a "natural" setting. Modern ideology often presents this experience as a physical relaxation of the body, but if we pay attention, it is first and foremost a relaxation of the mind, which lets go of its hold on the body and lets itself go into a communion, conscious or not, with the natural world around us. Suddenly, the stars, clouds, mountains, seas, and rivers, and the trees, flowers, insects, earthworms, and birds are with us again. We stop thinking that we are different from them; we first feel that we are all part of the same movement of the cosmos and of life. They are part of us, and we are part of them. Note that this is not an intellectual belief; it is a feeling rooted in our body, which may or may not find its way to consciousness, depending on how open we are to consciously listening to our feelings.

Sensing the world around us without any preconception is the shortest way to reconnect with outer nature. Not only does it allow us to be *in touch with* nature, it also allows us to be *touched by* nature, both sensorially and emotionally. As Abram (1996, 68) rightly points out, "To touch the coarse skin of a tree is thus, at the same time, to experience one's own tactility, to feel oneself touched *by* the tree. And to see the world is also, at the same time, to experience oneself as visible, to feel oneself *seen*. (. . .) We can perceive things at all only because we ourselves are entirely a part of the sensible world that we perceive! We might as well say that we are organs of this world, flesh of its flesh, and that the world is perceiving itself *through* us."

Being touched by nature is a gift that can transform our whole life. For we then stop feeling alone when we are alone; we feel connected to life that teems all around us, we belong to something bigger than us. Modernity has sought to cut humans off from nature by rejecting it outside them in order to better domesticate and dominate it. As ecologist Neil Evernden (1992, 116) points out, "Through our conceptual domestication of nature, we extinguish wild otherness even in the imagination. As a consequence, we are effectively alone, and must build our world solely of human artifact. The more we come to dwell in an explained world, a world of uniformity and regularity, a world without the possibility of miracles, the less we are able to encounter anything but ourselves." We crucially need others who are different from us to forge our own identity, both as individuals and as a species. And we need these others, not as mere objects of intellectual knowledge, but as active subjects who tell us about their lives and who touch us, physically and emotionally.

Although connection to life might look like an abstract concept to those who have not experienced it, it is a powerful feeling that responds to one of our most fundamental needs as sentient and empathic mammals. The evolution of the human species, like that of its pre-human ancestors, took place in a natural setting; the innate foundations of our behavior are therefore necessarily adapted to this natural setting. Since humans are empathic animals that spontaneously care for others—in particular, their young—and need care from others, it makes perfect sense from an evolutionary perspective that they would show "biophilia," that is, an innate propensity to seek connections with nature and other life forms (Wilson 1984).

Darwin (2011 [1871], 69, 79) already recognized that empathy (which was called "sympathy" in his time) has very likely played a critical role in the evolution of the human species: "In however complex a manner this feeling may have originated, as it is one of high importance to all those animals which aid and defend one another, it will have been increased through natural selection; for those communities, which included the greatest number of the most sympathetic members, would flourish best, and rear the greatest number of offspring." He then went on to explain how empathy within the human species naturally spread to other species: "Sympathy beyond the confines of man, that is, humanity to the lower animals, seems to be one of the latest moral acquisitions. (. . .) This virtue, one of the noblest with which man is endowed, seems to arise incidentally from our sympathies becoming more tender and more widely diffused, until they are extended to all sentient beings. As soon as this virtue is honoured and practised by some few men, it spreads through instruction and example to the young, and eventually becomes incorporated in public opinion."

In this passage, Darwin shows remarkable foresight in anticipating a trend that has been growing in recent decades, namely public concern for animal welfare and nature conservation. Where he was wrong was in his belief that this moral acquisition is recent and unique to humans. Caring for living beings from other species now appears to be a widespread, albeit uncommon, behavior among mammals. We also have many anecdotal examples of animals from other species coming to the rescue of humans in the wild. Thus, extending empathy to other species is by no means a human exception.

The bonds that are established between beings of different species can even be surprisingly strong. I experienced this myself a few years ago when I rescued a baby owl who had fallen, or was expelled by his parents, from his nest and was obviously calling for help. The baby owl, who still had his eyes closed, immediately adopted me as his new mother, as attested by the way he snuggled into my hand. "Little Moon" (*Petite Lune* in French) was the name I gave him. Little Moon soon became the center of gravity of the family as my wife, my youngest son, and I spent most of our summer taking care of him—we even took him with us on holiday abroad. Although we had no previous experience of rearing a bird, communication between Little Moon and us was so clear and intense that we were soon able to understand when he was happy, when he wanted to cuddle, when he wanted to play, when he wanted to poop, when he wanted to eat, and even when he wanted to eat a mouse rather than mealworms. After a few months, Little Moon became a young adult and decided he wanted to marry me, so he tried to pull me into his favorite shelters (that is how I first discovered he was a male). As I was unable to respond favorably to his marriage proposal, he wisely chose a young female owl as mate and came back to introduce us to her—with relatively limited success, as she was scared of us and remained at a reasonable distance. Eventually both of them were chased away by Little Moon's parents and moved to a better place.

I guess many people could probably recount similar experiences with other animals. The reason why I am telling this story here is that it provides a concrete illustration of how strong the connection to non-human life—especially untamed wildlife—can be and how important it is to our lives as humans. The ties that bound us together, Little Moon and I, were so strong that the few months we spent together remain one of the most powerful experiences of my life. I have been a biologist at heart since I was a child and I have observed and raised countless animals of all kinds, and yet the strength of our connection took me completely by surprise. I was touched by this little owl to the depths of my heart, which, despite my biological upbringing, I did not expect from a bird. Little Moon allowed me to rediscover in myself the power of immediate communication, without any intervention of thought and speech, which we are born with but are then educated to ignore and forget. Pepperberg (2009) reports a similar experience with her famous parrot Alex. We modern humans are used to thinking too much; we often forget our body, our sensations, our emotions, and our inner powers. Little Moon awakened me to some of these powers. He was to me a gift that fell from the sky and then returned to the sky.

Young children tend to be spontaneously "biophilic." When they have the opportunity to be in touch with nature and they are not deterred from doing so by either parental prohibitions or previous traumatic experiences, most young children show an almost unlimited curiosity and attraction to all living beings. The way their innate biophilia evolves later depends a lot on how they are educated and how much they are exposed to nature during their childhood and adolescence. Modern education is strongly focused on acquiring intellectual knowledge to the detriment of a real lived knowledge of oneself and the world. Therefore, it tends to cut us off from nature and from our spontaneous affinity for it, thereby relegating biophilia to the background, or even turning it into a pathological biophobia in adulthood.

The current fast-growing trend toward urbanization of the world's human population is another powerful factor that contributes to disconnecting people from nature, as most major cities in the world are highly artificial environments that only allow for an extremely limited direct experience of nature. Today, more than 55% of people live in urban areas, and this proportion is expected to increase to 68% by 2050. The gradual extinction of the experience of nature that results from modern education and urbanization generates a vicious spiral: once people are physically disconnected from nature, they tend to lose their emotional connection to nature as well and thus devalue it. This devaluation of nature in turn contributes to legitimizing and facilitating destructive individual and collective practices toward nature, which further increases human disconnection from nature. The end result of this vicious spiral may be a complete estrangement of people from nature, which can only be broken if people regain opportunities for meaningful interactions with the natural world (Pyle 1993; Soga & Gaston 2016). Unfortunately, scientific studies that have measured trends in experience of nature are still too few to fully assess the extent and generality of this process of extinction of experience (Cazalis et al. 2023).

The disconnection of modern humans from nature does not reduce their fundamental need for connection to life—simply, like several other fundamental needs in modern society, this need is then unmet, which generates conscious or unconscious frustration, the search for compensation in other activities, and even psychological dysfunctions of varying severity, known as "nature-deficit disorders" (Louv 2005). The discovery of nature-deficit disorders has recently led to a proliferation of new therapeutic approaches based on contact with animals. Many animals, in particular dogs, cats, horses, and dolphins, are now being used to cure a wide range of psychological dysfunctions and diseases in humans. This state of affairs is particularly ironic, as modern society is still largely based on the worldview developed by Descartes, who considered animals to be soulless automatons.

A growing number of scientific studies have sought to quantify "human-nature connectedness," that is, the degree to which humans feel connected to nature, and its causes and consequences. In a recent global meta-analysis of these studies, a few of my colleagues and I have found that both exposure to nature and mindfulness practices significantly enhance human-nature connectedness (Barragan-Jason et al. 2022). Mindfulness is simply a mental state in which a person maintains a moment-by-moment awareness of his or her thoughts, feelings, bodily sensations, and surrounding environment, thus focusing awareness on the present moment rather than dwelling on the past or imagining the future, as our minds do most of the time without us even noticing it. In turn, people who feel connected to nature are, on average, happier and healthier than people who do not (Capaldi et al. 2014); they also place a higher value on the integrity of both natural environments and human social communities. Thus, very simple changes in human behavior, such as practicing mindfulness and outdoor activities that increase contact with nature, might help to break the vicious spiral of disconnection from nature, and instead generate a virtuous spiral of reconnection with nature, which, interestingly enough, tends to favor at once human welfare and nature conservation.

Embracing nature as a fully animated world of which we are part has clear benefits, but it may also have some costs. For a world that is filled with active subjects of all kinds may be less easy to manage than an inert material world in which only humans count. Traditional modern ethics places all humans on an equal footing, which requires a social contract that guarantees enough equality and freedom for all people to trust the collective norms and institutions that regulate social life. Philosopher Michel Serres (1992) argued that the social contract that governs human social interactions should now be replaced with a natural contract that governs the interactions between humans and nature. Indeed, since the various components of what we call "nature" should be regarded as philosophical subjects, they should also be treated as subjects of law with rights and duties, like humans.

This, however, begs an important question: how could the multitude of non-human living beings possibly be represented in our collective norms and institutions without rendering these inoperative? First note that extending the concept of subject of law to natural entities does not mean that all entities should be equal before the law since they are qualitatively different by nature. Like values and ethics, law is a human construct; therefore, it is people who must decide how best to regulate their attitude toward the rest of nature. We saw in Chapter 6 that, although values and ethics are necessarily anthropogenic, they are not necessarily anthropocentric. Exactly the same applies to law and other social norms and institutions. Second, contemporary law has already been extended to sentient domestic animals without making law inoperative; extending it further to other natural entities is a simple matter of redefining our collective attitude toward these entities. In turn, granting legal rights to nature can be a powerful lever for changing our attitude toward nature (Boyd 2017).

Lastly, the most difficult aspect of law and justice is not so much to define the acceptable rules of conduct as to manage the conflicts that arise from the application of these rules. Modern law and justice are based on assessing what is "good" and "evil" and punishing "evil." The whole approach is a legacy of class-divided societies and the religions that accompany them: "good" is what corresponds to the collective fictions of the ruling class; "evil" is often what corresponds to the unmet needs of dominated classes. This approach is based on the use of force, which makes sense when it comes to maintaining the domination of one class over the others, but it is ineffective when it comes to managing conflicts between people in order to resolve them peacefully, and it makes no sense at all when it comes to managing conflicts with other species or natural entities. We will see in Chapter 8 that conflicts between people can be resolved in a much more effective and satisfying manner through approaches that seek to identify and meet the needs of the various people in conflict. Similarly, resolving conflicts between human and non-human entities is best achieved by ensuring that the inherent movements or needs of the various entities can be expressed. If we choose to extend our social norms and institutions to non-human entities, it is to learn to live with them, not to obstruct them.

Imagining what a world would be like where non-human beings are fully animated and taken into account for what they are, just as we are, seems difficult today because we have become so accustomed to the modern fiction that only humans are subjects that deserve consideration and respect. Yet the alternative view of a fully animated non-human world, which may seem so strange to modern adults, is extremely simple and natural because it corresponds to what we spontaneously felt and perceived when we were children. We probably have a lot to learn from animist hunter-gatherer people who have preserved this spontaneity, not in order to take over animism for ourselves, but in order to rediscover in ourselves the general movement of life that flows through us and unites us with all the animate and animate beings around us. Our deep connection with nature and the cosmos is the most essential thing for us humans, but we have forgotten this.

8

Recovering nature in us through our fundamental needs

The relationship of humans with nature is also the relationship of humans with themselves. For, whether we like it or not, nature is within each of us, in the physical form of matter and energy, in the chemical form of complex and diverse molecules, in the biological form of even more complex and diverse structures, cells, and organs, in the physiological form of bodily and mental processes, and in the psychological form of instincts, predispositions, and needs. The idea that humans can emancipate themselves from nature and dominate it is in truth a very curious idea, which is more a matter of fantasy than of reason, even though modern rationalism has made it its leitmotiv. It is enough to simply observe ourselves, without preconception and without complacency—or, which is sometimes easier, to observe our fellow human beings: we are indeed living beings in flesh and blood like others, which leads us to behave in ways that abstract reason alone cannot explain.

In fact, most of what we do on a daily basis has no other meaning than to participate in the general movement of life, without any particular logic or reason. Thus, we feel, we move, we eat, we breathe, we sweat, we defecate, we procreate, and we die in ways that are quite usual for a living being. It is only in our mind that we imagine that we are different from the animal kingdom, but, as we recalled in the first chapter, this is an optical illusion that makes us see ourselves as greater than we are. Where we really stand out from the rest of the living world and excel, however, is in our ability to use our mind to tell stories and create fictions. These fictions can become so powerful that they come to substitute for outer nature, which explains why we can give ourselves the illusion that we are emancipating ourselves from this outer nature. But outer nature is no different from our inner nature: it is made up of the same physical, chemical, and biological elements and forces, and these largely ignore the boundaries of our body to make the inner and the outer constantly communicate.

This little reminder of the elementary conditions of human existence is not intended to discredit us as a species, but rather to credit us with what we really are, namely living beings that participate in the general movement of life, and

Nature That Makes Us Human. Michel Loreau, Oxford University Press. © Oxford University Press 2023.
DOI: 10.1093/oso/9780197628430.003.0009

thus of nature. The sickly desire of modernity to emancipate itself from nature cuts us off not only from outer nature, but also from our inner nature, from what makes us living human beings. For centuries, Christianity has repudiated the human body and its animal impulses in order to glorify an immaterial soul free from all bodily defilement. While modernity has resolved to adopt a more neutral attitude toward the body, it has, on the other hand, pushed to the limit the exaltation of human freedom against a purely mechanical nature. In fact, there is nothing special about the body in the modern worldview: as an integral part of mechanical nature, the body obeys the same laws as all other components of nature. Nothing can be expected of it; we should learn to dominate it, even to modify it, in order to put it at the service of the Promethean program for the development of human power.

Recovering nature is not only about connecting with the non-human world around us, it is also about connecting with our own body, which embodies within us that world that Western civilization has cast out as an alien "nature." This is why many contemporary approaches to personal development begin with a very simple step consisting of connecting with one's body—all its parts and all the sensations that manifest themselves in the present moment. For many people, this reconnection to their body is a true revelation, which says a lot about the extent to which the collective fictions we adhere to, voluntarily or involuntarily, have a hold on our lives. Reconnecting to one's body and sensations is so powerful that it is a source of healing for many psychological wounds. This is the principle of new therapeutic approaches such as the "TIPI emotional regulation" technique (https://tipi.org), which allows people to uproot the fears that are at the root of recurrent or persistent psychological disorders by simply focusing attention on and embracing the bodily sensations associated with these fears. Mindfulness practices, which consist of intentionally focusing attention on the experience of the present moment without any judgment, also have the effect of reconnecting to one's body. It is perhaps not surprising, then, that these practices go hand in hand with a sense of connection to nature, as we saw in the previous chapter.

But there is more to human nature than just the body as traditionally conceived. One of our most enduring collective fictions, which has run through many societies since the dawn of humanity, is the separation of body and mind. This separation is probably rooted in the immemorial experience of dreaming, which seems to temporarily detach the mind from the body and let it float outside its material limits. We now know that dreaming is only an activity of the body. The fact that dreaming or thinking leads us to wander away from the physical limits of the body says nothing about the nature of this body. For the body itself never ceases to move beyond its physical limits to interact with the outside world, sometimes at great distances, by projecting itself into it through movement, sound, or odor, or by bringing it to itself through smell, hearing, or sight. Modern biology teaches us that, contrary to Descartes's philosophical presupposition, the body is by no means made of inert matter; it is living matter and activity that develop coherently in a delimited space. There is nothing in this definition to distinguish between what is "body" and what is "mind": the body, which is supposed to be material, engages in incessant immaterial activities; the mind, which is supposed to be immaterial, is so closely linked to the bodily matter that it often translates the state of the body into thoughts and, in turn, has a strong influence on it. Everything that happens in our body affects our mind, and everything that happens in our mind affects our body. The so-called psychosomatic illnesses illustrate this principle, but in reality, all illnesses, like all our activities, are psychosomatic in the sense that they combine body and mind. It must therefore be said that body and mind are only two arbitrarily separate facets of a single biological and physical reality.

What, then, is human nature, which is not reduced to either body or mind? This question has haunted modern philosophy since its inception because a coherent answer to this question cannot be given based on the presuppositions of modernity. Indeed, if humans are distinguished from the rest of the world by their immaterial soul, human nature should logically be defined by this characteristic. But this is impossible since nature is precisely defined as the material world from which human spirituality frees itself. The human material body cannot be used to define human nature either, since this would reduce humans to the material side from which they should detach themselves. Consequently, human nature has remained an insoluble enigma for modernity. The culmination of the insurmountable contradictions that accompany the dualism inherent in the modern worldview is probably found in the existentialist philosophy that flourished after the Second World War. Existentialism proclaimed loud and clear the fundamental freedom of choice of humans, as if they were pure spirits, free from all material fetters. In so doing, it consecrated the existential anguish and unhappiness of modern humans, alone with their conscience in front of a purely mechanical and absurd material world. As philosopher Erazim Kohak (1984, 4) forcefully noted, in *Being and Nothingness*, the famous work of existentialist philosopher Jean-Paul Sartre,

the human is *l'être-pour-soi*, the intentional, meaning-creating project wholly discontinuous from and in a fundamental conflict with the sheer, meaningless mass of what simply is, as *l'être-en-soi*. The human as a moral subject—"man," in the terminology of the age—is said to have no "nature": the ideas of "humanity" and "freedom" and the idea of "nature" appear fundamentally contradictory. The human here is a nothingness, a "godlike," arbitrary freedom to whom—or to which—nature, dead, meaningless, material, is at best irrelevant and typically threatening, to be conquered by an act of the will.

By contrast, if one accepts the inescapable fact that body and mind are one, the fiction of the modern human alone and free dissipates effortlessly, allowing human nature to come to light. Human nature is what defines humans' identity as a species, that is, what is given to every human being regardless of gender, origin, or culture, and what makes social life possible despite the many differences among individuals and cultures. The biological character-istics of the human body necessarily fall within the scope of this definition, but they are not the only ones; the psychological motivations common to all human beings, which weave the social bonds between them, must also be in-cluded. Therefore, human nature is also defined by the *fundamental human needs*, which include the needs of the body and mind of every human being.

What is it about? Like human nature, the concept of human need has been much discussed in philosophy and social sciences, but not much clarity has been achieved about it (Doyal & Gough 1991). Indeed, the modern worldview is incompatible with the very notion of fundamental human need because it asserts the supremacy of spirit over matter, of mind over body, of subject over object, and thus of free will over any form of natural constraints. In this view, there can be no objective, absolute needs dictated by nature, but only subjec-tive, relative needs, or wants, defined by the preferences of individuals and the cultural context in which they live. These wants are seen as being essentially unlimited, as are the differences between them in different cultures and histor-ical periods. This is why nearly all modern ideological and political currents, both left and right, reject the existence of universal, fundamental human needs. Modernity's intrinsic relativism in human affairs, however, is internally inconsistent because it provides no objective basis for understanding indi-vidual preferences or for making any moral judgements about individual or collective actions (Doyal & Gough 1991). In reality, modern ideology speaks more about the needs of capital as a collective fiction of humanity than about the real needs of flesh and blood humans. As abstract social wealth that aims at its own growth, capital is indeed based on humans' belief that their needs are unlimited and that they must consume more and more products and re-sources to increase social wealth.

Contrary to this modern belief, in-depth studies by a handful of psychologists and economists over the past century have shown that humans have a limited number of universal fundamental human needs (Maslow 1954; Max-Neef 1991; Rosenberg 2015). What varies greatly over time and across cultures are the specific satisfiers or strategies that people use to seek to satisfy these fundamental needs. The distinction between needs and strategies is difficult to make on the basis of purely intellectual considerations, which explains the widespread confusion about fundamental human needs in both the scientific and political literature. Unfortunately, this confusion is the source of many misunderstandings, psychological disorders, and interpersonal conflicts. In particular, by failing to clearly distinguish between their fundamental needs and the strategies they use to seek to satisfy them, and by consciously or unconsciously focusing their attention on their strategies, people often reproduce, without realizing it, strategies that are inherited from the past but which are no longer appropriate or effective in trying to satisfy unmet fundamental needs. In so doing, they become prisoners of repetitive behaviors that systematically set them up for failure or conflict in their relations with others or with themselves.

Practical experience, however, shows that people can spontaneously identify their fundamental needs when they step out of their intellectual judgments and connect to their body sensations and feelings. As we will see in the next chapter, paradoxically the body often speaks more clearly and more truthfully than the mind. Based on many years of practical experience in social and personal development, economist Manfred Max-Neef (1991) and psychologist Marshall Rosenberg (2015) have independently found that, beyond their many nuances, fundamental human needs can be grouped into a few coherent families of needs. In Max-Neef's (1991) classification, these families are as follows:

(1) subsistence (shelter, air, food, hydration, light, rest, reproduction . . .);
(2) protection (trust, harmony, peace, emotional and material security, support . . .);
(3) affection (love, belonging, empathy, sharing, respect . . .);
(4) understanding (clarity, communion, meaning, significance, unity . . .);
(5) participation (contributing to the well-being of self and other, cooperation, connection . . .);
(6) idleness (relaxation, play, laughter . . .);
(7) creation (beauty, evolution, expression, self-fulfillment . . .);
(8) identity (authenticity, self-confidence, respect for oneself and others . . .);
(9) freedom (autonomy, independence, sovereignty . . .).

Any classification is of course open to debate; in particular, some of the families listed above may partially overlap. Nevertheless, one cannot fail to be struck by the fact that the fundamental human needs identified so far extend far beyond the physiological or subsistence needs on which the focus has traditionally been placed. Only the first two families (subsistence and protection) include physiological and utilitarian needs, which underlines the importance of non-physiological and non-utilitarian needs in the daily life of human beings. This observation drawn from practical experience demonstrates the inanity of the separation of body and mind. In the Judeo-Christian and modern conceptions of the world, the body is supposed to be the expression of material constraints that restrict the exercise of human spirituality. But we now discover that all the virtues traditionally associated with spirituality, such as love, authenticity, self-fulfillment, and meaning, are inscribed in the body in the form of fundamental needs! Even human freedom, which modernity has emphasized so much in contrast to the alleged mechanistic determinism of nature, turns out to be an integral part of human nature. Practical experience further shows that the satisfaction or dissatisfaction of those needs which belong to what is traditionally called "mind" or "spirituality" generates extremely powerful bodily sensations and emotions, capable of affecting and orienting the whole life of individuals, and therefore also of social institutions. Human nature clearly does not know the boundary between body and mind.

Another striking aspect of the fundamental human needs listed above is that most of them evoke well-being and joy rather than misery and deprivation. The word "need" comes from the Old English word *nied*, which originally signified "force, violence, distress, anxiety, fear." Similarly, the French word for need, *besoin*, comes from an old word that originally signified "poverty, necessity," and later "work, worry." Even today, the notion of need retains a strongly negative connotation for most people, associated with misery and deprivation. This negative, depriving connotation comes straight from the Judeo-Christian worldview. In this view, need is what is natural and necessary to humans; it is the expression of their body, of their animality, of the non-human part of them from which they must distance themselves in order to reach the divine, the pure, the spiritual. Need is the hallmark of animals and of those whose social condition brings them close to animals, that is, the poor and the miserable. This conception has been perpetuated, practically unchanged, in modernity. Psychoanalysis, for example, which has played a considerable role in the historical development of human psychology, says no different. The father of psychoanalysis, Sigmund Freud, saw the human libido as an instinctive, primal, essentially animal force that drives humans to seek unlimited erotic pleasure from an early age. Society has to curb this instinctive "pleasure principle" in order to make orderly social interactions possible, by imposing a "reality principle" from above in the form of parental restrictions and laws. In developing this theory, Freud simply gave a new form to the old Judeo-Christian conception of human nature as being made up of dangerous animal needs, which the human mind and social institutions have to control and repress at all costs in order to attain true humanity.

The current of psychology known as "humanistic psychology," which developed in the United States in the mid-twentieth century under the impetus of Carl Rogers and Abraham Maslow, radically challenged this assumption. It demonstrated instead that, like their primate cousins and many other mammals, humans are spontaneously endowed with empathy and oriented toward life in society, and that the repression of their instincts has the exact opposite effect to that postulated by psychoanalysis, namely that of creating neurotic and asocial individuals. Maslow (1954) has probably best explained the fundamental error of psychoanalysis: instead of looking at healthy individuals, who could shed light on the psychological drivers of harmonious human development, Freud and the psychoanalysts took as their model the sick, severely dysfunctional individuals who usually came to them for treatment; they then claimed to apply this model to all humans. It is clear that such an approach can only lead to erroneous conclusions about the general development of the human personality. Maslow therefore chose the opposite path and devoted his work to the psychology of those individuals who showed the highest apparent degree of self-fulfillment.

His results are unambiguous: self-fulfillment results above all from the full satisfaction of our innate fundamental human needs. And his conclusions are radical:

> I might go as far as to say that sickness often consists of just *exactly* the loss of one's animal nature. The clearest specieshood and animality is thus paradoxically seen in the most spiritual, the most saintly and sagacious, the *most* (organismically) rational. (. . .) Healthy reason as definable today, and healthy instinctoid impulses point in the same direction and are *not* in opposition to each other in the healthy person (although they *may* be antagonistic in the unhealthy). (. . .) Their apparent antagonism is an artifact produced by an exclusive preoccupation with sick people. (Maslow 1954, 82, 84-85)

By opposing body and mind, animality and spirituality, Christianity and modernity have contributed to the creation of a human being who is certainly efficient in certain respects, but who is cut off from his own natural roots and deeply ill. Conversely, by reconciling body and mind, animality and spirituality, humanistic psychology has made it possible to rediscover a human being who is healthy in body and mind, and for whom the satisfaction of his or her needs is a source of self-fulfillment.

This new vision of humanity, oriented toward the fulfillment of all human capacities rather than the repression of a part of them, goes hand in hand with a radically different view of the notion of need. Just as a glass can be seen as half empty or half full, need can be seen from two opposing perspectives: that of unsatisfied need, linked to lack and poverty, or that of satisfied need, linked to fulfillment and flourishing. The first point of view is the traditional one of Christianity and modernity; the second is the new one outlined by humanistic psychology. It goes without saying that these two aspects of need are not incompatible, that every need presents itself, simultaneously or successively, in the form of lack and fulfillment, but practical experience shows that, for most people, the simple fact of adopting the point of view of fulfillment opens up a completely new field of vision which can go so far as to change their lives profoundly. For need is what connects us to the general movement of life that flows through us and transcends us. Recognizing, naming, accepting, and embracing our needs, experiencing their power and beauty, is a source of joy and richness, not of misery and poverty.

Rosenberg (2015), who helped develop this new vision through his experience as a therapist, used it to create a simple and practical but powerful communication process, called Non-Violent Communication (NVC), which is remarkably effective in resolving many psychological problems and interpersonal conflicts. Although there are many other similar approaches, the specific interest of NVC is, in my eyes, its explicit emphasis on the key role played by fundamental human needs. Indeed, the crucial step in NVC is the identification of the fundamental needs that underlie psychological problems or interpersonal conflicts, as these needs are often deeply hidden under a thick layer of beliefs and judgments, that is to say, of stories that people tell themselves about themselves and others. Once people's needs have been clearly identified and heard, Rosenberg claimed, based on a long experience in a wide variety of conditions—including armed conflicts—that a solution to a conflict is found spontaneously by the people involved in less than twenty minutes. The reason for this is simple: all human beings, regardless of their origin, culture, and education, share two essential things: the same fundamental needs and empathy, that is, the innate capacity, inscribed in the body, to experience and recognize the sensations, feelings, and needs of others (see Chapter 1).

Empathy alone is not sufficient to resolve conflicts between individuals or human groups as it can be channeled, during individual development, toward particular profiles of people. Some scientists have even suggested that empathy toward relatives may increase conflicts between human groups (Bruneau et al. 2017). In reality, it is not empathy that generates or exacerbates conflicts between individuals or human groups, but rather the *restriction* of empathy to particular profiles, which appear to us, rightly or wrongly, closer to our own. It is in this respect that fundamental human needs play a key role, as they help to remove the restrictions on empathy that have been put in place through education or previous experience.

As long as people talk about strategies and judgments, they remain in intellectual considerations that cannot arouse empathy and therefore cannot establish a deep connection between them. When they talk about their feelings and sensations, they can arouse empathy, but they do not yet give themselves and others the means to resolve the problems and conflicts that give rise to these feelings and sensations. On the other hand, when they are able to express their fundamental needs, any other person can spontaneously recognize these needs within him- or herself, understand the origin of the problem or conflict, and find solutions that satisfy the needs of all concerned. It is therefore access to our shared human nature, as it manifests itself through our fundamental needs, that allows us to connect or reconnect with each other.

Fundamental human needs thus allow us to reconnect with our inner nature. Do they also allow us to reconnect with outer nature? Existing classifications of fundamental human needs do not distinguish a specific need for connection with nature because this need is inextricably linked to all other needs. For instance, our subsistence needs clearly cannot be met without outer nature as the latter provides us with the physical, chemical, and biological conditions for our existence in the form of air, water, and food. Our close dependence on nature is equally, though perhaps less visibly, manifested in the satisfaction of our non-physiological needs. For instance, it is difficult, if not impossible, to imagine the full satisfaction of our needs for security, relaxation, empathy, connection, self-fulfillment, or meaning without any contact with nature.

The satisfaction of their fundamental human needs does not, therefore, cut people off from nature; on the contrary, it reconnects them to nature. When we listen to our bodies and our needs, nature ceases to be a foreign world from which we should free ourselves; it becomes our world again, the world of which we are part and which we embrace as such. The recognition of shared needs based on non-utilitarian interactions with other beings, both human and non-human, has profound consequences for our entire worldview. In particular, as we saw in Chapter 6, it destroys the separation between self-fulfillment and the fulfillment of the other, since the fulfillment of the other is the condition for my own fulfillment, and vice versa. The satisfaction of fundamental human needs thus naturally leads to the recognition of intrinsic values in the human and non-human world: others, human and non-human, are ends in themselves; they need me as much as I need them. Ethics and reason here merely extend and formalize the necessities of lived life.

Monotheistic religions and modernity have placed humans at the center of the world as they see it. But they have done so by cutting humans off from nature and their own nature, by rejecting their body and their needs, and by enslaving them to a spiritual ideal that is not of this world. Modern society has neither the aim nor the result of ensuring the satisfaction of fundamental human needs. On the contrary, it is a hindrance in this respect today, because it encloses people in an artificial world that distances them from their true nature, their true needs, and their true fulfillment. The full satisfaction of fundamental human needs requires the abandonment of the value system established by modernity and the development of a new worldview in which people accept to share with all non-human beings the qualities that they have unduly taken away from them.

9

Reunifying knowledge of body and mind

In order to reunify humans with nature, we must not only connect to nature outside, but also recover nature inside us, as we saw in the previous chapter. This means paying attention to our body and taking into account the precious information it carries in the form of sensations, feelings, and needs. Listening to and respecting the body, however, is incompatible with the philosophical foundations of modernity, which postulate that the body has nothing useful to teach us. One of the arguments that Descartes used to justify his famous postulate "I think, therefore I am" was that our body and senses can easily mislead us about what we are; only rational thinking gives humans the certainty of their real existence. For Descartes, as for the entire Judeo-Christian religious tradition, the body and its sensations, feelings, and needs are the expression of matter and nature in humans, which distance them from the spiritual and the divine and which must therefore be absolutely distrusted.

Of course, we have come a long way since Descartes. As a result of medical progress and the "sexual liberation" that has taken place over the past century, the idea that the body is fundamentally evil and repulsive has been greatly diminished, at least in the most economically "developed" countries. But neither sport nor sex is enough to get rid of the spirit-matter and subject-object dualities that are the foundation of modernity. Most citizens of the world's richest countries engage in physical activities to maintain their bodies and satisfy their need for sex, but it would not even occur to them that these physical activities could include anything spiritual or could be a source of knowledge. In other words, there is still a watertight partition between matter and spirit, between body and mind.

The same is true of the recent craze in some circles for everything to do with "spirituality," as opposed to the "materialism" of modern society. Many traditional spiritualist approaches perpetuate a long religious tradition that seeks to find human salvation in asceticism, in an iron discipline imposed on the body to reach the spirit outside the body, outside space, and outside time. In so doing, they consecrate the separation of body and mind that is the hallmark of both Christianity and modernity. As we saw in Chapter 4, capitalism has its

Nature That Makes Us Human. Michel Loreau, Oxford University Press. © Oxford University Press 2023.
DOI: 10.1093/oso/9780197628430.003.0010

origins in an ascetic Protestantism that sought salvation in labor as a means of giving glory to God on Earth. It is therefore somewhat paradoxical that spirituality is often seen as an antidote to the materialism of modern society, when the latter is a form of spiritualism in disguise that asserts the spiritual power of humans over matter. Where capitalism and spiritualism differ, however, is in the purpose of the human spiritual power: whereas capitalism is all about the Promethean exercise of the human spiritual power *over* matter, spiritualism is generally about the exercise of the human spiritual power *outside* matter. But both perpetuate the separation between matter and spirit, body and mind, that monotheistic religions and then modernity have built over two millennia.

In his book *Consilience*, Wilson (1998) made a vibrant plea for a reunification of human knowledge across the traditional divisions between the natural sciences, humanities, and arts. But the consilience he advocated was conceived as the realization of the Enlightenment ideal; it remained confined to the unification of intellectual forms of knowledge and did not challenge the separation of body and mind. Moreover, it was essentially based on the exercise of the scientific method, generalized to the entirety of our knowledge. Reunifying humans and nature requires an effort to reunify human knowledge that is more extensive and demanding than that. It is more fundamentally a question of reunifying all forms of knowledge, whether they come from what we are accustomed to calling "body" or from what we are accustomed to calling "mind." These two facets of our being, which we have arbitrarily separated, are sources of different types of information, and thus of knowledge of the world and of ourselves. Our body gives us access to an extraordinarily wide range of information about what is going on inside and outside us in the form of sensations, feelings, and needs, but we have become accustomed to deliberately ignoring this precious information and relying solely on information that has been sorted and filtered by rational critical thinking.

Rational thinking, which is the foundation of modern science, is an extremely valuable skill of the human mind as it allows us to sort and organize all our knowledge, both individually and collectively. I would even say that, although modernity strongly asserts human rationality, it does not give it the place it could and should have in our lives. Indeed, modernity invites us to apply rational thinking to the outside world, but not to our own inner mental world, that is, to our thoughts themselves. If we pay serious attention, we will discover that our minds are cluttered with a frightening number of false beliefs about ourselves and others. Underneath many of our seemingly rational judgments are inner voices from a distant past (often from a painful childhood experience) that tell us things like: "I will never make it"; "I am worthless"; or "I do not deserve to be loved." These are just a few examples of common false beliefs that each of us holds secretly but which affect our behavior throughout our lives—and the more we ignore them, the more they rule our lives. Perhaps the most tragic aspect of this is that everyone has more or less the same core beliefs, but each person believes that they apply only to him or her (or to the person next to them, which is often more comfortable). Critically examining our innermost beliefs would undoubtedly be one of the most useful applications of rational thinking. What applies to beliefs about ourselves applies, of course, also to the many judgments we make about the outside world, which we firmly believe without having bothered to examine their truthfulness carefully. A good example of how our innermost beliefs can be questioned and deconstructed using critical thinking is a process designed by Byron Katie called "The Work" (Katie & Mitchell 2003). This process invites us to ask a few very simple questions about our beliefs, and the answers to these questions often come as a complete surprise; they generally lead us to deeply change the way we view ourselves and others.

To be effective, however, rational critical thinking requires maintaining a strong connection to the body and the sensations, feelings and needs that are expressed in the body. While Descartes believed that our body and senses could easily mislead us, the truth is that our thoughts mislead us incomparably more often. Not only can we trust our sensations, feelings, and needs, but it is even the only way to achieve genuine rational thinking. Modern rationalism requires us to achieve rationality by rising above our body, but recent research in neurosciences shows that, in the absence of integration of the information provided by our sensations and feelings—which can happen as a result of certain brain lesions—thought is unable to lead us to make rational choices because it has no objective basis for evaluating the relevance of the various possible choices (Damasio 1994). It is the set of sensations, feelings, and fundamental needs inherited from our evolutionary history as a species that forms the biological basis for our rational choices and morality (Hauser 2006). Thus, the irony is that modern rationalism could not exist without this biological foundation, even as it claims to deny it. More importantly, the denial of our biological foundation has profound detrimental consequences for our mental health. When rational thought seeks to cut itself off from the body, it becomes an obstacle to connecting with ourselves and others, and thus it prevents us from satisfying one of our most fundamental human needs, generating a wide range of psychological dysfunctions. Body and mind are one. Depriving ourselves of the knowledge provided by the body condemns us to cut ourselves off from our own nature and perpetuate the domination of spirit over matter, which is the ultimate source of the current ecological crisis.

A number of contemporary approaches to spirituality begin with a connection to bodily sensations. This connection to the body is sometimes interpreted as a way of calming the body down in order to achieve a spirituality detached from bodily constraints, but the concrete reality of these "spiritual" experiences, as I have been able to live and observe them myself, is quite different: what is experienced is an opening of the body to the outside world, a connection with the outside world that goes beyond the boundaries of the physical body, but which nevertheless remains deeply anchored in bodily feelings. It is as if the body were mind and the mind, body. In other words, far from manifesting the domination of mind over body, of spirituality over materiality, this type of experience opens awareness to the deep unity of body and mind, of materiality and spirituality. There is nothing mysterious about these experiences: it is enough, in any circumstance, to pause and bring attention to one's physical sensations, feelings, and thoughts in the present moment—that is, to practice mindfulness—to feel a deep connection with the rest of the world. The difficulty is not in feeling this connection, which is present all the time whether we want it or not and which we can embrace without any effort; the difficulty, in our modern hyperactive and hypermental world, is in the act of pausing, of stopping the incessant flow of thoughts and beliefs that run through our minds and protect us from our deepest fears, to simply welcome what is here and now.

The duality of body and mind is what has led modern civilization to assert the domination of mind over body, of spirit over matter, and of humans over nature. Conversely, the unity of body and mind can enable us to rediscover harmony between body and mind, between spirit and matter, and between humans and nature. This unity, however, can only exist and be maintained in us and for us if it exists in our consciousness. This is why the unity of knowledge is so important. As long as we ignore the body as the primary source of knowledge, we cannot prevent our mind from reasserting its supremacy over the body and re-establishing the duality of body and mind. We will persist in the illusion that the body needs a mind to master it and nature needs humans to dominate it and give it meaning. Cognitive dualism underlies all the other dualisms that Western civilization has created since the Neolithic revolution. Although I have always been a scientist by both profession and passion, I cannot conceive of science and reason as the only forms of knowledge. Science illuminates the world and reason remains reasonable only insofar as they recognize their limitations and rely on intuitive forms of knowledge that come from what we call the body. Only by finding or rediscovering the unity of these different forms of knowledge can we find our place and have a just action in the world.

Kohak (1984, 32-33) had this nice metaphor to describe the complementarity between our different forms of knowledge: "Daylight, with its individuating brightness and its pressing demands, is the time of *technê*. (. . .) Nighttime, by contrast, is the time of *poiêsis*, (. . .) of deep dreams. (. . .) Dusk is the time of philosophy. (. . .) It is at dusk that humans can perceive the moral sense of life and the rightness of the seasons." I would add that the body and its sensations are of all times, they run through life from beginning to end—without them we would not be, we would have access neither to science, nor to poetry, nor to philosophy. We need to learn again to embrace and celebrate life that flows within and around us, from birth to death, in all its forms and rhythms. We need to stop glorifying knowledge derived from abstract thought alone and rediscover the unity of our knowledge and identity as a species, which connects us to the general movement of life.

As soon as we embrace life, we re-establish the connection with what we call nature. For nature is nothing else than the movement of life that we have pretended to push out of us. We *are* nature, as is everything around us. We are made of the same elementary particles, the same atoms, the same molecules, the same cells, and the same organs as the other living beings. Our body is a complex ecosystem that contains many more bacterial cells than human cells. We are constantly traversed by flows of energy and matter that also pass through all the beings, living or non-living, with which we interact. Whether we consciously accept it or not, we are permanently in communication, even in physical communion, with all these beings. This is the inescapable reality revealed to us by both modern physics and ecology. It is therefore not a question of denying modern science, but rather of taking it seriously, and linking it to what we experience and feel in our body. Our body is our first gateway to nature.

Bringing together all the knowledge of body and mind and achieving full awareness of the self and of the world is both very simple and very difficult. It is very simple because it is accessible to any person at any time: it is enough to embrace life as it manifests itself inside and outside us. But it is very difficult because, to do so, we have to stop the incessant flow of thoughts and beliefs that we have inherited, not only from our own individual history, but also from the history of humanity, which is perpetuated in our social environment in the form of powerful institutions and collective fictions. Whether we are aware of it or not, many of our actions are motivated by fear—fear of death, fear of loneliness, fear of not being good enough, fear of meaning nothing to others, fear of lacking love, and so on. The personal fictions that we create for ourselves in childhood and the collective fictions maintained by social institutions serve, to a large extent, to protect us from our deepest fears and thus avoid being overwhelmed by painful emotions. This is why it is so difficult, at first, to let go of the thoughts that we consciously or unconsciously dwell on. For the emptiness of thought can open the door to joy and serenity, but it can also open the door to deep feelings that we have not yet had the opportunity or capacity to face and embrace. Once we are adults, only through a long and arduous, if exhilarating, introspection and inner healing work can we put our fears in their rightful place, embrace them as expressions of precious parts of our identity, and thus open the way to a full awareness of ourselves and the world.

Direct contact with nature is of course another gateway to nature and to a more unified knowledge, which we have already explored in Chapter 7. Knowledge of nature comes first from simply enjoying the presence of the inanimate and living beings we encounter and the life they bring out in us. This direct connection can then be extended through intellectual learning about what they are, what they tell us about the universe, about life, and about ourselves. The sciences of the universe, of life, and of the human being are only the conscious and systematic extension of the unified elementary knowledge that each of us spontaneously acquires through contact with the beings that populate nature.

The unity of humans and nature, of body and mind, and of the various forms of knowledge, is spontaneously present in young children, albeit in purely intuitive forms. It is to a large extent the education they receive from their parents, from school, and from society at large that makes them forget this original unity and instills in them the beliefs and institutions of modernity, including the dichotomy of humans and nature, of body and mind. Young children also have cruder fears and less effective mechanisms to protect themselves from them because their cerebral cortex is less developed and, as a result, these fears and protections have not yet had time to develop into beliefs firmly anchored in their body and consciousness. They are therefore better able to maintain their connection to nature as long as they are encouraged to do so. This is why it will probably be easier to work toward restoring the unity of humans and nature by completely rethinking the education of children, who will make the world of tomorrow, than by seeking to transform adults, even though educating, or re-educating, adults is also essential, if only to maintain connection with their children.

Modern psychology argues that young children, like so-called primitive peoples, are spontaneously animistic, i.e., they attribute an intention or a "soul" to objects and non-human living beings (Piaget 2013 [1926]). This childish animism would reflect the young child's confusion between him- or herself and the external environment and would constitute an obstacle to establishing causal relationships in the events he or she experiences. Consequently, one of the roles of modern education is to establish a clear distinction between self and external environment, and between cause and effect. There is no doubt that distinguishing between what belongs to me and what belongs to others, and between what is cause and what is effect, is useful and necessary for the development of the human personality and thought—indeed, these distinctions play a fundamental role both in the identification of one's social responsibility and in scientific thinking. At the same time, however, one cannot fail to notice that these distinctions are closely linked to the founding myths of modernity, in particular to the affirmation of the human individual as the sole thinking subject in the face of a soulless world governed by mechanical relations of cause and effect. Thoughts are like most of the substances we ingest: in small doses they are often useful or necessary, but in too large a dose they become toxic. Between the recognition of my identity in the world and the proclamation of the human individual as the only thinking subject in the world, between the recognition of causal relationships in the world and the assertion that the world is a soulless machine, there is a gulf that nothing in the real world justifies.

All modern education is based on the fundamental assumption shared by both monotheistic religions and modernity that the specificity of humans is to be thinking subjects whose duty is to rise above their material body. Therefore, human beings cannot rely on their body, their sensations, their feelings, and their innate needs; the human mind must be shaped by a painful learning process of denial and discipline. Although discipline has been greatly relaxed in public schools over the past fifty years or so, education in many countries continues to be conceived fundamentally as a systematic learning process of intellectual knowledge that will enable the child to become, in adulthood, the thinking and working subject that modernity has defined. In this process, physical exercise is necessary insofar as it allows the body to give way to the development of the mind, free from bodily constraints. While physical and psychological violence is much less present today than it was in the past, the discipline of intellectual learning has remained largely intact.

If we want to get out of the fundamental contradictions of modernity and stop destroying nature, we need to rethink education completely. Education should be conceived rather as the process by which human beings develop the set of talents and skills that enable them to satisfy their fundamental needs and contribute to the flourishing of life in and around them (Rosenberg 2003). Thus, education cannot be separated from life; it is a process of self-development that enables each person to flourish life within and without. Knowledge is not to be found outside, ready to be ingested.

Knowledge is first and foremost knowledge of and by oneself—of one's body, one's sensations, one's feelings, one's needs, one's place in the world, how to care for oneself and for others, whether human or non-human. The intellectual knowledge accumulated by society then makes it possible to broaden this self-knowledge, but it can in no way replace it, otherwise we will lose the unity of the self and the coherence of our action in the world. Self-knowledge radiates into the world, but knowledge of the world does not necessarily radiate into the self.

The forms that this educational process can take are extremely varied. Numerous attempts, more or less successful, have been made over the past century to set up so-called alternative schools and personal development approaches that are inspired, to varying degrees, by the principle presented above. It is not my intention to evaluate them or to make concrete proposals here because I believe, generally speaking, that what is important in life is to have a clear awareness of the goal to be achieved. I then trust that life will find a thousand ways to achieve this goal concretely.

Trusting one's body, welcoming all the valuable information it provides, using one's reason in a reasonable way, and then integrating all the knowledge, whatever its source, into a coherent whole that guides action: these are some of the basic conditions that allow each human being to take care, with benevolence and discernment, of oneself, of others, and of everything we call nature. Thinking divorced from our bodily sensations, feelings, and needs takes us away from nature and ourselves; thinking connected to our bodily sensations, feelings, and needs helps us to be fully ourselves in nature. Life does not know the boundary between body and mind; we also have to get rid of this artificial boundary in our consciousness if we are to contribute to the flourishing of life inside and outside us.

10

Building a social and economic order that serves life

Perhaps the most formidable obstacle to restoring the unity of humans and nature is the present global social and economic order. Capital has penetrated nearly all exchange relations between human beings on Earth; it therefore decides on the material survival of the vast majority. As we have seen, capital is based on an extraordinarily powerful set of collective fictions that diverts people from the satisfaction of their natural needs and leads them in a headlong rush to the unlimited growth of abstract social wealth in the form of money. As long as people remain subject to these collective fictions created by thousands of years of history, there is little chance that they will be able to re-establish a more peaceful and harmonious relationship with the rest of nature.

Admittedly, capital can perfectly well accommodate new constraints linked to changes in consumer behavior or laws enacted by political authorities in favor of taking better account of the degradation of natural systems. Much of this adjustment occurs spontaneously through the creation of new enterprises or new production techniques, as recent history abundantly attests. Over the past fifty years or so, many countries have implemented more restrictive legal frameworks to limit the environmental impact of certain production activities. At the same time, consumer behavior has evolved toward a greater consideration of product quality, including the impact of products on health and the environment, at least in the most "developed" countries. New companies and new production techniques have emerged in response to these changes.

But the critical question is this: have these changes led to a reduction in the ecological footprint of humanity, or at least of those countries that have most adjusted their production to meet these new constraints? Not in the least, unfortunately. On the contrary, greenhouse gas emissions and threats to biodiversity have continued to rise, reaching new records year after year. Only the recent Covid-19 pandemic has been able to temporarily (and to a very limited extent) reduce some of the adverse effects of the human economy on natural systems in recent history, for the simple reason that it was a disturbance outside the economic system. Capital is a highly dynamic system, but there is no evidence to

Nature That Makes Us Human. Michel Loreau, Oxford University Press. © Oxford University Press 2023.
DOI: 10.1093/oso/9780197628430.003.0011

support the widespread belief that it could lead to a new equilibrium between humans and nature. When technological advances or restrictive public policies result in improved efficiency in the use of natural resources, the lower cost of resource use often generates an increase in demand for these resources. The rebound effect of consumption then cancels out the positive effect of resource use efficiency, ultimately leading to the total amount of resources used being maintained, or even increased. This paradoxical result is well known in environmental economics as the Jevons paradox. More fundamentally, as I have argued in previous chapters, capital is, in essence, a process of accumulation of abstract wealth, and thus only external factors can set a limit to this process as it has no intrinsic internal limit. It seems therefore futile to expect it to evolve spontaneously toward a new balance between humans and nature. Its inability to do so is not the result of a technical defect; it is the very essence of capital that makes it unfit for this purpose. An economic order based on abstract wealth and relative scarcity is simply inadequate to deal with the physical and biological constraints of a finite Earth system.

If capital does not do the job, what social and economic order will? The truth is that there is currently no coherent alternative to the existing social and economic order, which explains the recurrent difficulties of "green" political currents to convince in the long run. Communism once claimed to play this role, but the totalitarian excesses of the so-called communist regimes in Stalinist Russia, Maoist China, and elsewhere have profoundly undermined confidence in this alternative. Much could be said about these regimes, which have little in common with communism as the early utopians and Marx imagined it, but this is not my purpose. It suffices for me to point out here that the original communist ideal, conceived as the political and economic form of a united human community, responds neither to the current challenge of building a new relationship between humans and nature, nor to the complex reality of both human and non-human nature. Biology, ecology, psychology, and social sciences now show us clearly that cooperation and competition are inextricably intertwined at all levels of life; therefore, an ideal focused on cooperation alone is not in tune with real life and its multiple challenges.

Developing a new ideal model of economic organization, as the early utopians did, is probably not the most urgent task today either. On the one hand, I do not believe that a model of an ideal society is particularly useful to guide thinking and action; it could even easily become an obstacle if reality does not conform to it. On the other hand, the creation of a new social and economic order is an eminently collective process, which no single individual could summarize or anticipate. What seems essential to me is to identify some general principles that outline the goal toward which human society should strive in order to create and maintain the unity of humans and nature, and thus a balanced relationship with the rest of nature. I trust life and collective creativity to find the technical means to reach this goal.

If any social and economic order is to be sustainable in the long run, it must ensure that the conditions for the flourishing of life are maintained, both within human society and in the natural systems on which it depends. This imperative means in particular the satisfaction of fundamental human needs. As we saw in Chapter 8, these needs are the expression of nature in us; their satisfaction is what allows us to be part of nature in the same way as all other living beings. The degree of development of human societies should therefore be measured, first and foremost, by their capacity to satisfy fundamental human needs, and thereby to allow the flourishing of both human life and its environment. This was the conclusion that Max-Neef (1991) also reached in his study on the conditions for the economic development of contemporary societies, which led him to promote a "human-scale development." This principle may seem obvious, yet human societies have nearly systematically deviated from it for thousands of years. Primitive hunter-gatherer societies did in fact respect this principle because they were content to ensure the growth, survival, and social life of their members without further pretension. But the deviation from this elementary principle has been growing ever since the Neolithic revolution, culminating today in the subjection of almost all of humanity to a principle of abstract social wealth accumulation. It is true that the accumulation of wealth goes hand in hand with the satisfaction of many of the fundamental needs of the privileged fringe of society, but it exceeds it by a wide margin, while at the same time it does not ensure the satisfaction of the most elementary needs of a significant part of society, because this is simply not the purpose of economic activity.

If we accept the basic principle that the economy is about satisfying fundamental human needs rather than creating and distributing wealth, then the whole modern social and economic edifice has to be rethought. First of all, an important feature of fundamental human needs is that they are limited in number but can be satisfied in many different ways. Experience further shows that the more we are aware of our fundamental needs, the easier it is to find a way to satisfy them, especially by accessing our inner resources. The latter are much more important than most people realize, because contemporary society has accustomed us to frantically searching outside, in the consumption of new products, for the treasures that we often possess inside ourselves. A society based on the satisfaction of fundamental needs is the exact opposite of a straitjacket, because it allows the fulfillment of all, individually and collectively. But the paradox is that once we are fulfilled, we discover that self-fulfillment requires relatively little, despite the fact that making new experiences is an integral part of our fundamental needs. As long as our material security is assured, our connection to nature, to others, and to ourselves is often what matters most, and it does not necessarily require great means. Self-fulfillment is not incompatible with the "happy sobriety" advocated by the pioneer of agroecology Pierre Rabhi (2010) as long as this "sobriety" stems naturally from the full satisfaction of our needs and not from a voluntary asceticism that distances us from our needs. When we are truly fulfilled, we discover that the most precious things are simple things, within everyone's reach. By discovering our fundamental needs, we rediscover what is truly precious to us. And what is truly precious to us is not measured in euros or dollars, nor in the quantity of possessions.

If affluence is defined by the adequacy of available resources to satisfy fundamental human needs rather than by the quantity of goods produced, then "primitive" hunter-gatherer societies were paradoxically closer to true affluence than modern society is. This is indeed the astonishing conclusion that Sahlins (2017 [1972], 1-2) drew from his study of hunter-gatherer societies:

> By the common understanding, an affluent society is one in which all the people's material wants are easily satisfied. To assert that the hunters are affluent is to deny that the human condition is an ordained tragedy, with man the prisoner at hard labor of a perpetual disparity between his unlimited wants and his insufficient means. For there are two possible courses to affluence. Wants may be "easily satisfied" either by producing much or desiring little. The familiar conception, the Galbraithean way, makes assumptions peculiarly appropriate to market economies: that man's wants are great, not to say infinite, whereas his means are limited, although improvable: thus, the gap between means and ends can be narrowed by industrial productivity, at least to the point that "urgent goods" become plentiful. But there is also a Zen road to affluence, departing from premises somewhat different from our own: that human material wants are finite and few, and technical means unchanging but on the whole adequate. Adopting the Zen strategy, a people can enjoy an unparalleled material plenty—with a low standard of living. This, I think, describes the hunters. And it helps explain some of their more curious economic behavior: their "prodigality" for example—the inclination to consume at once all stocks on hand, as if they had it made. Free from market obsessions of scarcity, hunters' economic propensities may be more consistently predicated on abundance than our own.

A society based on the satisfaction of fundamental human needs can thus both ensure abundance for all and produce relatively little by comparison with modern society. The "Zen road to affluence" is the road to the flourishing of both human and non-human life on Earth.

An economy geared to satisfying fundamental human needs implies a "human-scale development" as advocated by Max-Neef (1991), and thus downscaling many economic projects and enterprises compared with the current situation. Sometimes great things can be achieved through large collective projects that would not be possible without the pooling of ideas, resources, and work of many people. A society based on the satisfaction of fundamental human needs is entirely compatible with large-scale projects as long as these respond to a real common aspiration. But many of the large-scale projects that have punctuated human history since the Neolithic revolution have been expressions of the Promethean excesses of ruling classes, rather than of a consideration of human needs. Was the Great Wall of China, for example, worth the suffering of the hundreds of thousands of soldiers, peasants, and prisoners mobilized for its construction? Was the conquest of space in the last century worth the slow destruction of our own planet? "Small is beautiful," proclaimed economist Ernest Schumacher (2010) in the title of a now famous book. Satisfying fundamental human needs implies that most social and economic activities should be designed at the size of what a human being is adapted to, that is, a community of a few dozen or a few hundred people. This means either relatively small projects and enterprises, or, in the case of larger projects and enterprises, organization into sufficiently independent subsets so that each person can find a satisfactory place in them.

Large cities currently concentrate most of humanity's economic and cultural activities, and they are growing steadily as people leave the countryside. Less than a third of the world's human population used to live in cities in 1950, more than half live there today, and more than two-thirds will do so by 2050 if current trends continue. But large cities also concentrate human misery: about a billion people live in extreme poverty, especially in unsanitary slums, and most people are disconnected from nature and their own needs. Cities, too, are an expression of the Promethean excess that has permeated all social activity, while the countryside is increasingly transformed into giant agricultural enterprises to feed the urban population. Both cities and countryside are plagued by the scourge of gigantism, which is one of the hallmarks of modernity. There is an urgent need to redefine the occupation of space in such a way that both humans and other living species can find their place and rediscover the links between them. The satisfaction of fundamental human needs and the fulfillment of individuals, society, and nature do not depend primarily on pharaonic projects and considerable resources, but rather on a social organization that privileges the connection to others and to nature, from the local to the global scale.

The same applies to the organization of labor and production. In a society geared toward the satisfaction of fundamental human needs, production is no longer motivated by the acquisition of wealth or profit, but by the satisfaction of the needs of the producers and consumers of the goods and services produced. The only criterion governing the growth or decline of a company or an economic sector is the quantity of goods and services necessary to satisfy these needs. This means, among other things, that entire sectors of the contemporary economy would no longer be relevant. For example, the luxury production, advertising, or armaments sectors, as well as a large part of public and private administration, which are currently over-bloated but do not contribute to satisfying fundamental human needs, could be largely dismantled in order to reinforce other, more useful branches of activity.

On the other hand, the distinction between work and creative or leisure activity would become blurred, as is already the case today in certain branches of activity that call largely on workers' creativity, such as artistic creation and scientific research. Reallocating labor toward sectors that contribute to satisfying fundamental human needs should make it possible to distribute the most unpleasant tasks among a greater number of people, and thus to reduce the amount of time each of them devotes to these tasks. The nature of work itself would change as everyone would have a clear awareness of the contribution they make. Instead of being experienced as a constraint to which one must submit in order to ensure one's survival and that of one's family, work would be transformed into a means of participating in social activity and contributing to collective well-being, as has been the case in many experiences of community living in recent or ancient history.

The creation of new projects and enterprises often depends on the vision, energy, and charisma of one or a few people. It is therefore important to give everyone the opportunity and the means to achieve this. But once a project or company reaches a certain size, collective organization becomes the main factor in its development. The concentration of power in the hands of one or a few people is neither useful nor effective; on the contrary, the free participation of everyone and shared governance are the best guarantees that a project or company reaches its full potential (Laloux 2015). Insofar as it is still necessary, remuneration for work should also be compatible with the objective of ensuring the participation of everyone and the satisfaction of their fundamental needs. Concretely, this means a drastic reduction in the income inequalities that have been growing steadily over the past decades.

Remuneration or income means money or currency. It is legitimate, however, to ask whether money is compatible with a society based on the satisfaction of fundamental human needs. Indeed, money is the form that abstract social wealth takes in capitalism because it can be accumulated without limit and can be exchanged for any other commodity. Does not keeping money mean keeping the door open to the accumulation of capital and to social relations motivated by profit, which are incompatible with the goal of satisfying human needs? To answer this question, it is important to bear in mind that money existed long before capital. Money as currency is a general means of exchange that allows goods and services to be exchanged between different members of a society when there is division of labor. Many primitive forms of money arose from exchanges between communities and individuals long before rare metals took over and became organized and state-guaranteed means of exchange. Any system of exchange involving large numbers of people requires some form of currency.

It is hard to imagine a society as large and complex as the contemporary global human society doing without any form of exchange between its billions of members, unless a gigantic centralized planning system be put in place, in comparison to which the five-year plans of former Soviet Russia would pale. Even if such planning were made technically possible by a highly sophisticated computerized system, it is difficult to see what advantage it would have over an exchange system, as it would likely have much less flexibility and an equally limited degree of control. The maintenance of an exchange system, and therefore of money, seems inevitable, at least initially, to ensure the complex metabolism of human society on a large scale.

To ensure that the economy is geared toward the satisfaction of fundamental human needs and not toward the accumulation of wealth, it is not necessary to eradicate money, but rather to collectively redefine the contours of its power of action. Society is free to establish not only a range of variation in labor income, but also a range of variation in the amount of money that a person or a company may store, an average or maximum lifetime of monetary symbols, or any other measure necessary to maintain the compatibility of the use of money with the general goal of satisfying fundamental human needs. It can also act on the creation and destruction of money to maintain an overall quantity of money in circulation compatible with the capacity of natural systems to renew themselves and provide a constant flow of natural resources necessary for economic activity, as proposed, for example, by the *Ex Naturae* nongovernmental organization (https://exnaturae.ong). Money would then cease to be a collective fiction inherited from the past that weighs on our consciences and actions like a leaden blanket and bears the stigma of thousands of years of domination of humans over other humans and over nature. By redefining its properties and use, society could make it a means of circulating goods and services adapted to its new objectives, and modify it according to its changing needs.

Redefining the goal and organization of the global human economy already seems a daunting task. But perhaps the most difficult task in establishing a new relationship between humans and nature is to drastically reduce people's collective footprint on the biosphere. Humanity's ecological footprint is mainly governed by two interdependent factors: the size of the human population (all other things being equal, the more people there are, the higher their collective footprint), and the average amount of natural resources used per capita (all other things being equal, the more natural resources a person uses, directly or indirectly, the higher his or her individual footprint). It is difficult to assess precisely how the transition from a society based on the accumulation of abstract wealth like ours to one based on the satisfaction of fundamental human needs would affect the average amount of natural resources used per capita. The latter is likely to fall sharply in the so-called developed countries, which enjoy a high average standard of living despite high social inequalities, but it is unclear how it might change in the so-called developing countries, where a large proportion of the population lives in poverty. Taking the "Zen road to affluence," however, should make it possible to substantially reduce the global average per capita ecological footprint, which has been growing very rapidly for over a century.

Satisfying fundamental human needs should also contribute to reducing the size of the human population in the long run. The particularly rapid growth of the world's human population over the past two centuries can be explained by several factors, but all these factors are related, directly or indirectly, to the tremendous development of the productive forces and social inequalities in modern capitalism. On the one hand, the expansion of industry and services has required a corresponding expansion of labor, especially cheap labor in the less "developed" countries. On the other hand, the human birth rate typically follows a bell-shaped curve with respect to access to resources: it rises from a state of starvation, when individuals have access to so few resources that their reproduction is severely limited, to an intermediate state of limited access to resources, characteristic of the poorest contemporary countries where abundant offspring are a way to reduce economic insecurity, and then declines again when individuals have access to abundant resources, as is the case in the richest contemporary countries, where the birth rate has declined sharply in recent decades (Henderson & Loreau 2019). Economic inequality between the world's richest and poorest countries has been a powerful driver of global human population growth over the past two centuries, as it has kept much of humanity in a state of limited access to resources, which tends to push up birth rates, while attracting the labor force created by these high birth rates to the almost continuously expanding industry and services, largely in rich countries. Unequal access to resources and differences in technological progress across the world's regions are a major threat to the sustainability of contemporary human societies (Henderson & Loreau 2021). Professional optimists have repeatedly told us that the demographic transition that has taken place in the world's richest countries over the past decades will solve all the problems associated with human population growth by itself, but this optimistic forecast is based more on a blind faith in the myth of progress than on a rigorous examination of recent trends. Of course, the demographic transition will continue, but the pace of its effects on the size of the world's human population is likely to be much slower than the pace of ongoing changes in the global climate and in the biosphere, as the world's human population is projected to continue to grow throughout the twenty-first century (Gerland et al. 2014).

A society based on the satisfaction of fundamental human needs would no longer provide the motives for such a population growth. By giving everyone the necessary access to resources, by promoting individual fulfillment and by drastically reducing social inequalities, it would provide the conditions for a gradual, voluntary decline in the size of the human population, which is essential to reduce humanity's ecological footprint and thus to achieve a sustainable balance with the biosphere. Note, however, that demographic adjustments take time, such that they would not be enough to reach a sustainable balance with the biosphere for a fair amount of time, even if a change in society were to take place immediately.

Hunter-gatherer communities ensured their demographic balance, and thus their balance with their natural resources, by a rather strict control of births and deaths due to a combination of natural and cultural factors. Their nomadic way of life led them to travel regularly over long distances to find new hunting and gathering grounds. These long journeys were difficult for pregnant women, young children, and old people in poor health, so these communities often had a set of customs to limit births and facilitate the death of sick old people. A future society based on the satisfaction of human needs would no longer have the same incentives to limit births because, since the emergence of agriculture, nomadism has given way to a sedentary lifestyle. But the profound security that would come from meeting everyone's fundamental needs and the freedom of choice that would come from widespread access to education and contraception would be powerful incentives for spontaneous birth control. The recent evolution of the most "developed" contemporary societies shows that birth limitation arises spontaneously under conditions that fully respect individual freedom.

The conscious and voluntary acceptance of the limits of the human lifespan is another condition for achieving a sustainable balance of the global human population with the biosphere. In our current fast-paced, growth-oriented society, death is seen as an abomination. This abominable view of death seems to have developed following the Neolithic revolution and the emergence of religions. All religions hold out, in one form or another, the promise of overcoming time and death through the immortality of the soul. The thinkers of modernity then transposed this immortality of the soul into earthly life by proclaiming that progress would gradually lead to the immortality of humans themselves. The myth of progress is a very effective collective fiction, which allows humans not only to spur the accumulation of wealth, as we have seen in Chapter 5, but also to cheat on death and imagine the realization of a divine existence on Earth. Religions and modernity hate death because death reminds us that we are part of this perishable material life from which we must escape at all costs to reach the divine. Recovering nature in us therefore implies that we reclaim death as an integral part of life.

In order to be serene, the reappropriation of death requires that people stop seeing themselves as isolated individuals. Death is fundamentally a moment of universal life that runs through humanity as it does through all living beings. We are not immortal as individuals, but we are immortal as a manifestation of life that links us to all living beings, past, present, and future. In this sense, and in this sense only, the immortality of the soul does exist: if the soul is the "breath of life" that manifests itself in us and not the individual property that Christianity and modernity have imagined, it continues to manifest itself in the beings around us and it connects us to all of life. The belief in the immortality of the soul is a distorted expression of the connection that genuinely unites us with life and nature. This link is the most precious thing in us. It is up to us to rediscover it and to cultivate it with passion.

To be sustainable, the human social and economic order must be conceived as serving life, not the other way around. The economy must become once again what it is supposed to be, which is, etymologically in Greek, "the management of the house." But the house we are dealing with today is that of humanity as a whole, that is, the biosphere and the Earth system as a whole. With the exception of a few minority currents in ecological economics, economics as a scientific discipline still devotes most of its efforts to finding the best ways to accumulate abstract social wealth. If it is to rise to the challenge of helping to manage our global common house, economics must stop focusing exclusively on the exchanges of goods, services, and money between humans; it must also be concerned, first and foremost, with the exchanges between humans and the ecosystems on which they depend and which they inexorably transform. What we need today is a genuine ecology of the human species, an integrative understanding of the meaning and limits of human action within nature, so that life on Earth, including that of the human species, can flourish and be perpetuated.

11
Embracing life that flows through us

In Part I of this book, I showed that, far from being an accident, the current global ecological crisis has its roots in 11,000 years of human history and is inextricably linked to the philosophical and economic foundations of modern Western civilization, which rules the world today. This means that we have no historical example to draw on to overcome a crisis of this magnitude. Although some societies have been able to successfully face environmental problems throughout history, these have been small societies facing relatively localized problems. The scale, speed, and complexity of current changes in climate, biodiversity, ecosystems, the biosphere as a whole, and human societies themselves are beyond the scope of those experienced by our species since its inception.

Given the scope and depth of the current global ecological crisis, technical solutions will not suffice. While technical solutions will of course be needed to address each of the many facets of this crisis, it is a pernicious illusion to believe that a set of practical recipes will be enough to overcome it. On the one hand, the major changes we are facing today are closely interconnected, so that they cannot be broken down into a set of isolated problems that could be solved independently. On the other hand, these changes are not the result of technical flaws; they are the result of the worldview conveyed by modernity, which itself is based on a set of powerful collective fictions built up bit by bit and passed on over thousands of years by hundreds of generations of men and women who, for the most part, were not even aware of them. Some of the practical solutions envisaged or implemented today to mitigate the harmful effects of the extension of human activities on natural systems are clearly necessary. For instance, nature reserves are necessary to preserve what is left of biodiversity on the planet, although there is debate about how they are designed and implemented. In contrast, others are a continuation of the technological excesses of modern civilization and have every chance of accentuating the ecological crisis rather than solving it. This is the case, for example, with the new approaches proposed by geo-engineering, which aims to manipulate the Earth system to slow or stop global warming, or transhumanism, which aims to transform human beings to increase their physical and mental capacities. However laudable the intentions behind these ideologies may be,

Nature That Makes Us Human. Michel Loreau, Oxford University Press. © Oxford University Press 2023.
DOI: 10.1093/oso/9780197628430.003.0012

they perpetuate the headlong rush characteristic of modernity toward ever deeper and rapid, and therefore increasingly uncontrolled, transformation of nature and of people themselves.

In the face of the current global ecological crisis, all existing ideologies are powerless or insufficient. Monotheistic religions and modern philosophical and economic thought have largely contributed to generating the current crisis by glorifying human domination over nature; it is pointless to expect them to renounce their very foundations. Marxism has produced a relevant critique of some aspects of capitalism as an economic system, but it has retained from the modern philosophical tradition its glorification of human domination over nature, which prevents it from offering a credible alternative to overcome the ecological crisis. Conversely, "deep ecology" has produced a relevant critique of the Promethean program of human domination over nature, but it has not yet succeeded in proposing a coherent vision of a new world freed from the vitiated foundations of modernity and capitalism.

Nor is there any hope in the past. In particular, there can be no return to a mythical Golden Age, which never existed. While the hunter-gatherer way of life that prevailed in prehistoric times had a number of advantages, including the relative stability of human populations and their interactions with the rest of nature, it is incompatible with the current level of human development, both in terms of the qualitative development of the human personality and social interactions and in terms of the quantitative number of human beings on the planet today. The hunter-gatherer way of life is only viable for small populations living on territories large enough to ensure the natural renewal of all their resources without human intervention. The generalization of the hunter-gatherer lifestyle to the entirety of humanity today could only be conceived at the price of a drastic reduction in the size of the world's human population. Such a perspective does not offer a solution to the current ecological crisis.

It is not to the past that we must turn, but to the future. It is up to the present generations to build a new relationship with nature that allows the human species to flourish within a flourishing nature. This means abandoning the persistent myth of the separation between humans and nature and consciously reintroducing people into nature, within the biodiversity of which they are a part. This challenge requires people to rethink almost everything they have been used to, from their existential aspirations to the form and content of the contemporary global economy.

This challenge, unprecedented in the history of humanity, seems disproportionate, almost inconceivable. And yet it is well within our reach. It is not so much the goal as its implementation that poses a problem. In Part II of this book, I have shown that a profound transformation of our relationship with nature and our way of life is, in fact, quite simple to achieve. It is a matter of turning our attention to the foundations of human life and of life itself and drawing all the consequences. Reconnecting to nature and to our own nature is not intrinsically difficult. We all have this connection spontaneously at birth; it persists in our bodies throughout our lives and is only waiting to be revived at any time. It remains buried deep within us and disappears only in our thoughts, to which we have developed the bad habit of giving an inordinate power in both the Judeo-Christian and modern traditions. The obsession with divine perfection, which invisibly permeates all of modern society, constantly pushes us out of the earthly, the corporeal, the sensitive, the temporal, the material, and thus it violently represses this connection from our consciousness. By giving voice to our body and the sensations, feelings, and needs it manifests, we can easily rediscover our connection to the world, to nature, to life, to the cosmos. For, despite the collective fictions of modernity that we have been fed since childhood, we are, profoundly and fundamentally, living beings, inseparably linked to the general movement of life and thus to all of nature. In a way, it is only a matter of letting go and allowing nature to express itself within us, without effort.

A society based on the satisfaction of fundamental human needs and the flourishing of nature and human beings within it is not difficult to conceive either. Utopians have imagined many versions of such a society throughout history, although their understanding of fundamental human needs and the deep connection the latter establish between humans and nature was rudimentary at best. As I showed in the previous chapter, the economic principles governing such a society are compatible with the technical means developed by modern capitalism, although the purpose of economic activity is fundamentally different. There is no need to imagine returning to the living conditions of the mythical caveman (who also never existed, since prehistoric hunter-gatherers painted but did not live in caves) to overcome the current ecological crisis. Rather, it is in the blossoming of human creativity in the service of a rediscovered connection with nature that we must seek the path to a profound transformation of modern society. The utopias of recent decades, such as the one sketched by writer Ernest Callenbach (1990) in his novel *Ecotopia*, make extensive use of individual creativity and freedom to imagine a world in which humans have rediscovered their natural simplicity while making sensible use of modern technology.

The main obstacle on the way to a profound transformation of contemporary society lies in the power of the collective fictions that Western civilization has constructed over the last eleven millennia and which today permeate the thinking and behavior of the vast majority of human beings, to the point where they seem almost natural and unalterable. The thing about collective fictions is that they are collective; they are built up gradually, generation after generation, in the course of a society's development. The larger the community and the older the fictions, the more difficult it is to free oneself from them, because the entire social fabric, including the conditions for the material survival of its members, has been impregnated with them for centuries, even millennia. This is why social systems tend to persevere in their essential characteristics to the end of their potential; then, when this potential is exhausted, they enter a phase of decline or even collapse. It is now known that the collapse of civilizations is a much more common phenomenon in history than previously thought (Diamond 2005).

If collective fictions play such an important role in human history, would not it be enough to create new collective fictions to transform the world, as proposed, for example, by film director Cyril Dion (2018)? There is no doubt that dreams play an indispensable role in any social transformation—without dreams of a better world, no social transformation is possible. Also, every society is cemented by a worldview shared by its members. The development of a new worldview is therefore essential if the society of tomorrow is to be built on new foundations. But a worldview and a shared dream are not arbitrary fictions created by an act of will. If they are to guide action, they must carry within them a new truth in line with the deep needs of people and the evolution of society. They must also be sufficiently anchored in people's practices and consciousness so that they do not become new beliefs which in turn hinder the transformation of the world. Unfortunately, there is no shortage of examples in human history of shared dreams becoming obstacles to the very transformations they announce. One need only think of the words of Jesus Christ announcing a universal human brotherhood which ended up being used to justify the bloodiest monarchies, or, more recently, the communist ideal of a free and egalitarian society which ended up being used to justify regimes of totalitarian oppression. Dreaming, imagination, and storytelling are essential human activities that allow us to project ourselves into a future that has not yet happened and to cement social bonds, but they are also sources of false beliefs that feed fear and paralyze action. We must therefore learn to dream while living in the present moment and, in order to do so, to make a clear distinction between fiction and reality, between narratives that carry us forward toward a greater truth and false beliefs that pull us back. In a way, we already need to begin the process of reunifying the different modes of knowledge that I outlined in Chapter 9.

Humanity's oldest collective fictions do not simply shape our consciousness; they also shape our behavior in deeper ways. The strongly hierarchical organization of societies since the Neolithic revolution is perpetuated from generation to generation not only through the collective fictions transmitted to children by their parents, by their school, and by society as a whole, but also through the traumas, small and large, that these collective fictions inflict on children and which then shape their behavior as adults. Although children are less and less physically abused at home and at school today, they continue to suffer intense, if less apparent, psychological distress as a result of the high expectations placed on them and the constant judgments they are subjected to at home and at school. Each of us has the imprint of the domination of humans over other humans and over nature in the inner wounds that we carry within us and that lead us to reproduce, in spite of ourselves, the same wounds in the next generation. It is these inner wounds that drive people to indulge in the addictions of modern society, such as sex, power, and money. In turn, these addictions lead them to perpetuate the foundations of the social and economic organization of society, and thus the resulting destruction of nature. This is why being aware of our inner wounds and healing them is so important: by healing our own wounds, we not only make our lives lighter and more beautiful, we also indirectly contribute to stopping the spread of these wounds, and thus of destructive behaviors toward others and nature, around us (Schwartz & Sweezy 2020). This is the same principle as vaccination: by getting vaccinated, not only do we protect ourselves from disease, but, more fundamentally, we help stop the spread of disease to others. And once a sufficient proportion of the population is vaccinated, the disease dies out by itself. The same goes for collective fictions and the wounds they inflict on us: once enough people get rid of them, they will disappear on their own, leaving room for the development of new, more fulfilling shared narratives and behaviors and a clearer, less distorted view of reality.

A growing number of people have already gained enough awareness and self-confidence to change their lives and engage in new forms of relationship with nature on a small scale. These experiences are part of a slow evolution of collective consciousness and behavior that heralds a more profound social transformation. But the current economic order, which carries within it the separation of humans and nature, is still making its mark on people's consciousness and actions on a daily basis and is a powerful brake on this transformation. For example, anyone wishing to change their lifestyle today must find a way to accommodate the constraints of market exchange and money, unless they completely remove themselves from the social fabric. These constraints are more or less strong depending on people's interests, talents, and skills, but it is undeniable that they are a deterrent for many. This is why only a minority of people who are sufficiently clear, motivated, and confident in their project dare to make a radical change of lifestyle today. The constraints are even stronger at the collective level. On the one hand, a collective change in the foundations of the existing social and economic order requires a consciousness and determination shared by a large majority of the population. On the other hand, the expansion of capital is largely based on credit and debt, so that the present generations seem to have their hands tied by the choices made by past generations. Debt is one of the pillars of modern capitalism, as it maintains the belief that only further growth of the economy would pay off the debt that financed past growth, thus perpetuating capital accumulation indefinitely. This is why, despite the growing awareness of the importance of the contemporary ecological crisis and its consequences, ecological concerns are still often overlooked when major political and economic choices are made.

History teaches us that major social transformations generally result from the convergence of two factors: on the one hand, a slow maturation of collective consciousness following a profound transformation of the relationships that people have with each other and with their environment, or the emergence of a powerful new social project; on the other hand, a situation of major crisis, which suddenly shakes the little confidence that people still have in the existing social and economic order. There is little doubt that the first factor is in the making today, but there is also little doubt that awareness of the importance and urgency of an in-depth transformation of our social relations and our relationship with nature is still far from having spread throughout society on a global scale. As for the second factor, we have just had a timid glimpse of what awaits us in the coming decades on the occasion of the global health crisis caused by the Covid-19 pandemic. Climate change and biodiversity loss, which continue to accelerate today, are bound to generate increasingly dramatic social and economic crises, whether through droughts, floods, food shortages, new diseases, wars, or massive influxes of migrants driven from their homes by deteriorating living conditions. It is therefore highly likely that the conditions for a profound social transformation will mature in the coming century.

These prospects, however, are far from rosy. They herald difficult times ahead for present and future generations. Moreover, they offer no guarantee that the social upheavals to come will be in a desirable direction. "Collapsologists" have recently popularized the idea of a coming collapse of modern civilization as a result of the increasing intertwining of ecological, social, and economic constraints in contemporary global society (Servigne

& Stevens 2015). Scientists themselves are beginning to seriously explore catastrophic worst-case scenarios of worldwide societal collapse, and even human extinction (Kemp et al. 2022). While consideration of worst-case scenarios is essential to fully appreciate the dangers humanity is likely to face in this century and the centuries to come, there is no evidence to support collapsologists' firm belief in the imminent collapse of modern capitalism. First, many others before them—for instance, the revolutionary Marxists of the early twentieth century—have predicted the imminent collapse of capitalism in the past. Second, the collapse of a society or a civilization is not an automatic phenomenon; it depends on the behavior of people. The consciousness and collective action of people often follow a slow and tortuous course, and it would be audacious indeed to pretend to understand and predict its intricacies.

The relatively long time taken by social transformations contrasts sharply with the relatively short time taken by ongoing natural transformations, whose pace is accelerating. Modernity has denied the elements of the natural world the status of active subject and has reduced them to a set of inert objects, constituting the static scenery of a human history that was supposed to take center stage in the theater of life. Is it not paradoxical that today it is this supposedly inert nature that takes center stage and calls on humans to wake up from their torpor? As Latour (2015, 99) rightly notes, "human societies seem to be resigned to playing the role of stupid object, while it is nature that unexpectedly takes on the role of active subject! (. . .) This is the meaning of the New Climate Regime: 'warming' is such that the old distance between background and foreground has melted: it is *human* history that seems cold and *natural* history that is taking on a frenetic pace."

This reversal of the scene of history not only puts to the test the philosophical presuppositions of modernity, which are becoming increasingly clear for what they are, namely fictions that conveniently justify unlimited exploitation of nature. It also puts to the test the capacity of humanity to respond to the accelerating changes of nature. For the natural forces that we have set in motion by burning fuels, clearing land, and consuming or exterminating plants and animals lavishly are slow and powerful forces that, once set in motion, are very difficult to stop. The climate system has such an inertia that it would take centuries, even millennia, to erase the trace of ongoing climate change. The evolution of life is even slower: while many threatened species could still be saved by changing our lifestyle and behavior today, once a mass extinction has occurred, it takes millions of years to recover the level of biodiversity that prevailed before. We might even be one of the last generations that can still avoid the disaster scenario of a runaway process of combined global warming and biodiversity loss that would change life on Earth for a very long period of time.

Thus, there is a considerable gap today between the urgency of the transformation of human society that would be necessary to avoid future natural and social disasters and the slowness of changes in human consciousness and collective action. Added to the inertia of human behavior and consciousness is the inertia of human population dynamics. Human demography typically responds to changes in human behavior with a delay of about one or a few generations because individual behavior is essentially shaped during childhood, while the effects of changes in behavior in terms of births and deaths are manifested in the adult state. Demographic inertia further lengthens the response of societies to environmental challenges, as the world's human population and thus its impact on humanity's ecological footprint are likely to continue to grow for several decades after any social changes have been made. Powerful natural and social forces are thus working together to darken the future of human societies and, with them, of much of life on Earth.

Time is running out: the longer the current social and economic order based on unlimited accumulation of abstract wealth and exploitation of nature continues, the faster humanity's ecological footprint will grow and the more difficult it will be to achieve a voluntary decline of the global human enterprise to fit the limits of the Earth system. If there is one fundamental thing that both cybernetics and ecology have taught us, it is that dynamic systems are characterized by feedbacks that take time to manifest themselves. For example, left on its own with no other limiting factor, a predator population tends to exploit its prey to the point where the prey runs out; only then does a decline in the predator population begin, and this decline continues long after the prey population begins to grow and replenish itself again. Humanity is in much the same position as a predator approaching the peak of its prey consumption: it is close to the point where the available resources provided by the biosphere will start to decline. But by the time resource availability declines, it is too late to reach a possible equilibrium point: the decline in the prey population in turn leads to an inexorable decline in the predator population, and this decline is all the more violent the longer the predator's behavioral and demographic responses are delayed.

One might object, of course, that humans do not fit the classic model of a predator exploiting a single population of prey, in particular because they use a wide range of different natural resources, they have a greater awareness than many other animals, and they have a great capacity for technical and technological innovation. But these assets all have their limits. For example, while the diversity of resources used by the human species is undoubtedly an important asset that allows it to absorb shocks due to fluctuations or depletion of some of them, it is precisely this diversity that is threatened today with biodiversity loss. Moreover, not all resources are substitutable; some resources that are essential to the functioning of modern society, such as fossil fuels, rare metals, and natural areas, are irreplaceable and risk being rapidly depleted. Similarly, human consciousness in principle gives us a great capacity to anticipate external constraints and to adjust our behavior to these constraints. But the superiority that our consciousness would confer on us has been largely overestimated, as I showed in the first chapter. In particular, the *collective* consciousness that would allow human societies to anticipate and respond collectively to a situation that calls into question the collective fictions on which these societies are based is much more limited than is generally thought. Finally, while technological innovation does enable humans to overcome some of the disadvantages of their previous tools and techniques, it has so far not contributed to reducing the total quantity of natural resources they use—quite the contrary.

Thus, the question today is not so much whether there will be a decline in the human population, its economic activity, and its ecological footprint in the centuries to come, but what form this decline will take and where it will lead the human species. Either humanity succeeds in changing its relationship with nature and reducing its collective ecological footprint in a deliberate, organized, and sufficiently rapid way, in which case its decline may be quantitative only, may be more or less gradual, and may lead to the qualitative flourishing of the human species as well as of many other species. Or it fails to do so, and its decline will take the far more painful form of famines, wars, epidemics, and perhaps even the extinction of the human species, among many other species.

Although the latter perspective may seem more likely every day in view of the current situation in which the anthropic pressure on the biosphere and on the climate system is accelerating while collective human action to end this pressure is glaringly insufficient, this is no reason to give up and contemplate, powerless, the prospect of a seemingly inevitable catastrophe. First, history is never played out in advance. On the contrary, history shows us that major crises can be the occasion for major upheavals, mobilizing people's energies toward new goals and new projects. Rather than deploying a treasure trove of natural resources and human energies toward such a futile goal as establishing a few human colonies on Mars—as some dream of today in a very real sense—these resources and energies could be redeployed toward the far more useful and fulfilling goal of caring for life on our own planet. Creativity is one of the most remarkable attributes of the human species: let us use it to give meaning and joy to our time on Earth!

Second, both demoralizing pessimism and blissful optimism are attitudes of flight that prevent us from accepting reality as it is. Both attitudes feed our bad habit of telling ourselves stories that take us away from real life, with all its opportunites and obstacles, and the joys and sorrows that come with them. Our most essential reality is that life flows through us and never ceases to surprise us. The future may look bleak outside, but nothing can silence the power of life in us. Contrary to the illusion conveyed by modernity, life is not at our service; we are an integral part of life. As such, we have the freedom to put ourselves at the service of life, and to do so with all the energy and joy that life gives us, whatever future it may hold. The widespread idea that joy, trust, and willpower make sense only as long as the future is worth it is yet another facet of the disembodied idealism that pervades Christianity and modernity. For the latter, life, body, and matter are nothing; only the ideal toward which they can lead us, before or after death, matters. This mortifying vision cuts us off from real life as it unfolds in and around us.

Reconnecting with and caring for nature is first and foremost reconnecting with life that flows through us and connects us to all living beings, past, present, and future. Life is not waiting for a better future: it is simply there, in every place and in every moment. It is this ever-present movement that awakens in us joy, trust, and willpower when we embrace it without restraint or ulterior motive. It is not by lamenting past, present, or future misfortunes that we will accomplish anything beautiful and useful, but rather by embracing life as it is, here and now. Thus the obvious becomes clear: life is the most precious thing we have. Life calls us to give the best of ourselves to work toward its fulfillment as an integral part of nature. What could be more beautiful and useful than to respond to this call?

References

Abram, D. (1996). *The spell of the sensuous: perception and language in a more-than-human world*. Vintage Books, New York.

Adams, W. M. (2004). *Against extinction: the story of conservation*. Earthscan, London.

Bacon, F. (2016 [1620]). *The new organon, or true directions concerning the interpretation of nature*. CreateSpace Independent Publishing Platform, Amazon.

Barragan-Jason, G., de Mazancourt, C., Parmesan, C., Singer, M., & Loreau, M. (2022). Human-nature connectedness as a pathway to sustainability: a global meta-analysis. *Conservation Letters*, 15, e12852.

Beyries, S., & Joulian, F. (1990). L'utilisation d'outils chez les animaux: chaînes opératoires et complexité technique. *Paléo*, 2, 17–26.

Bimbenet, E. (2011). *L'animal que je ne suis plus*. Gallimard, Paris.

Bohm, D. (1983). *Wholeness and the implicate order*. Ark Paperbacks, London.

Bookchin, M. (2010). *Une société à refaire: vers une écologie de la liberté*. Editions Ecosociété, Montréal.

Boyd, D. R. (2017). *The rights of nature: a legal revolution that could save the world*. ECW Press, Toronto, Canada.

Bradshaw, G. A., Schore, A. N., Brown, J. L., Poole, J. H., & Moss, C. J. (2005). Elephant breakdown. *Nature*, 433, 807.

Bruneau, E. G., Cikara, M., & Saxe, R. (2017). Parochial empathy predicts reduced altruism and the endorsement of passive harm. *Social Psychological and Personality Science*, 8, 934–942.

Callenbach, E. (1990). *Ecotopia: the notebooks and reports of William Nestion*. Bantam, New York.

Capaldi, C. A., Dopko, R. L., & Zelenski, J. M. (2014). The relationship between nature connectedness and happiness: a meta-analysis. *Frontiers in Psychology*, 5, 976.

Carruthers, P. (2009). Invertebrate concepts confront the generality constraint (and win). In: *The philosophy of animal minds* (ed. Robert W. Lurz). Cambridge University Press, Cambridge, pp. 89–107.

Cazalis, V., Loreau, M., & Barragan-Jason, G. (2023). A global synthesis of trends in human experience of nature. *Frontiers in Ecology and the Environment*, in press.

Currie, C. R., Scott, J. A., Summerbell, R. C., & Malloch, D. (1999). Fungus-growing ants use antibiotic-producing bacteria to control garden parasites. *Nature*, 398, 701–704.

Daly, H. E. (1977). *Steady-state economics: the economics of biophysical equilibrium and moral growth*. W. H. Freeman, San Francisco.

Damasio, A. R. (1994). *Descartes' error: emotion, reason, and the human brain*. G. P. Putnam's Sons, New York.

Darwin, C. (2011 [1871]). *The descent of man, and selection in relation to sex*. Pacific Publishing Studio, Seattle.

de Fontenay, E. (1998). *Le silence des bêtes: la philosophie à l'épreuve de l'animalité*. Fayard, Paris.

DeGrazia, D. (2009). Self-awareness in animals. In: *The philosophy of animal minds* (ed. Robert W. Lurz). Cambridge University Press, Cambridge, pp. 201–217.

Derrida, J. (2006). *L'animal que donc je suis*. Editions Galilée, Paris.

Descartes, R. (2021 [1641]). *Meditations on first philosophy*. Graphyco Editions, Amazon Italia.

Descola, P. (2005). *Par-delà nature et culture*. Gallimard, Paris.

de Waal, F. (2005). *Our inner ape*. Riverhead Books, New York.

de Waal, F. (2009). *The age of empathy: nature's lessons for a kinder society*. Harmony Books, New York.

Diamond, J. M. (1992). *The third chimpanzee: the evolution and future of the human animal*. Harper Collins, New York.

Diamond, J. M. (2005). *Collapse: how societies choose to fail or succeed*. Viking Penguin, New York.

Dion, C. (2018). *Petit manuel de résistance contemporaine: récits et stratégies pour transformer le monde*. Actes Sud, Arles.

Doyal, L., & Gough, I. (1991). *A theory of human need*. Macmillan Education, Houndmills, UK.

Ellis, B. (2002). *The philosophy of nature: a guide to the new essentialism*. McGill-Queen's University Press, Montreal and Kingston.

Engels, F. (2012 [1925]). *Dialectics of nature*. Wellred Books, London, UK.

Evernden, N. (1992). *The social creation of nature*. John Hopkins University Press, Baltimore, MD.

Evernden, N. (1993). *The natural alien: humankind and environment*. 2nd edition. University of Toronto Press, Toronto, Canada.

Ferry, L. (1992). *Le nouvel ordre écologique: l'arbre, l'animal et l'homme*. Bernard Grasset, Paris.

Flahaut, F. (2008). *Le crépuscule de Prométhée: contribution à une histoire de la démesure humaine*. Editions Fayard, Paris, France.

Foster, J. B. (2000). *Marx's ecology: materialism and nature*. Monthly Review Press, New York.

Freitag, M. (2008). *L'impasse de la globalisation: une histoire sociologique et philosophique du capitalisme*. Les Editions Ecosociété, Montréal.

Gerland, P., Raftery, A. E., Ševčíková, H., Li, N., Gu, D., Spoorenberg, T., et al. (2014). World population stabilization unlikely this century. *Science*, 346, 234–237.

Hadot, P. (2004). *Le voile d'Isis: essai sur l'histoire de l'idée de Nature*. Editions Gallimard, Paris.

Harari, Y. N. (2011). *Sapiens: a brief history of humankind*. Vintage, London.

Harari, Y. N. (2015). *Homo Deus: a brief history of tomorrow*. Harvill Secker, London.

Hauser, M. D. (2006). *Moral minds: how nature designed our universal sense of right and wrong*. Abacus, London.

Heidegger, M. (1971). *Nietzsche II*. Gallimard, Paris.

Henderson, K., & Loreau, M. (2019). An ecological theory of changing human population dynamics. *People and Nature*, 1, 31–43.

Henderson, K., & Loreau, M. (2021). Unequal access to resources undermines global sustainability. *Science of The Total Environment*, 763, 142981.

Howard, S. R., Avarguès-Weber, A., Garcia, J. E., Greentree, A. D., & Dyer, A. G. (2018). Numerical ordering of zero in honey bees. *Science*, 360, 1124–1126.

Janicaud, D. (2005). *La puissance du rationnel*. Gallimard, Paris.

Johnson, C. N., Balmford, A., Brook, B. W., Buettel, J. C., Galetti, M., Guangchun, L., et al. (2017). Biodiversity losses and conservation responses in the Anthropocene. *Science*, 356, 270–275.

Katie, B., & Mitchell, S. (2003). *Loving what is: four questions that can change your life*. Harmony Books, New York.

Kemp, L., Xu, C., Depledge, J., Ebi, K. L., Gibbins, G., Kohler, T. A., et al. (2022). Climate endgame: exploring catastrophic climate change scenarios. *Proceedings of the National Academy of Sciences of the USA*, 119, e2108146119.

Koch, P. L., & Barnosky, A. D. (2006). Late Quaternary extinctions: state of the debate. *Annual Reivew of Ecology, Evolution, and Systematics*, 37, 215–250.

Kohak, E. (1984). *The embers and the stars: a philosophical inquiry into the moral sense of nature*. University of Chicago Press, Chicago.

Krupenye, C., Kano, F., Hirata, S., Call, J., & Tomasello, M. (2016). Great apes anticipate that other individuals will act according to false beliefs. *Science*, 354, 110–114.

Latouche, S. (2004). *La Mégamachine: raison technoscientifique, raison économique et mythe du progrès.* Editions La Découverte, Paris.

Laloux, F. (2015). *Reinventing organizations: vers des communautés de travail inspirées.* Editions Diateino, Paris.

Latour, B. (2015). *Face à Gaïa: huit conférences sur le nouveau régime climatique.* Editions La Découverte, Paris.

Laurans, Y., Rankovic, A., Billé, R., Pirard, R., & Mermet, L. (2013). Use of ecosystem services economic valuation for decision making: questioning a literature blindspot. *Journal of Environmental Management*, 119, 208–219.

Loreau, M. (2010). *The challenges of biodiversity science.* Excellence in Ecology. International Ecology Institute, Oldendorf/Luhe.

Loreau, M. (2014). Reconciling utilitarian and non-utilitarian approaches to biodiversity conservation. *Ethics in Science and Environmental Politics*, 14, 27–32.

Loreau, M., Hector, A., & Isbell, F. (2022). *The ecological and societal consequences of biodiversity loss.* ISTE, London, and John Wiley & Sons, Hoboken.

Louv, R. (2005). *Last child in the woods: saving our children from nature-deficit disorder.* Algonquin Books, Chapel Hill, NC.

Louys, J., Braje, T. J., Chang, C.-H., Cosgrove, R., Fitzpatrick, S. M., Fujita, M., et al. (2021). No evidence for widespread island extinctions after Pleistocene hominin arrival. *Proceedings of the National Academy of Sciences of the USA*, 118, e2023005118.

Mancuso, S. (2018). *The revolutionary genius of plants: a new understanding of plant intelligence and behavior.* Atria Books, New York.

Maris, V. (2010). *Philosophie de la biodiversité: petite éthique pour une nature en péril.* Buchet/Chastel, Paris, France.

Marx, K. (1965 [1867]). *Le Capital, Livre I.* Bibliothèque de la Pléiade, Gallimard, Paris.

Maslow, A. (2006 [1971]). *Etre humain.* Eyrolles, Paris.

Maslow, A. H. (1954). *Motivation and personality.* Harper & Row, New York.

Mason, J. (2005). *An unnatural order: why we are destroying the planet and each other.* Lantern Books, New York.

Max-Neef, M. (1991). *Human scale development: conception, application and further reflections.* Apex Press, New York.

Mies, M., & Shiva, V. (2014). *Ecofeminism.* 2nd edition. Bloomsbury, New York, USA.

Millennium Ecosystem Assessment. (2005). *Ecosystems and human well-being: biodiversity synthesis.* Millennium Ecosystem Assessment. World Resources Institute, Washington, DC.

Morizot, B. (2018). *Sur la piste animale.* Actes Sud. Arles.

Mumford, L. (2010 [1934]). *Technics and civilization.* University of Chicago Press, Chicago.

Naredo, J. M. (2003). *La economia en evolucion: historia y perspectivas de las categorias basicas del pensamiento economico.* 3rd edition. Siglo XXI de Espana Editores, Madrid.

Parr, L. A. (2001). Cognitive and physiological markers of emotional awareness in chimpanzees (Pan troglodytes). *Animal Cognition*, 4, 223–229.

Pelluchon, C. (2011). *Eléments pour une éthique de la vulnérabilité: les hommes, les animaux, la nature.* Editions du Cerf, Paris.

Pepperberg, I. M. (2009). *Alex & me: how a scientist and a parrot discovered a hidden world of animal intelligence—and formed a deep bond in the process.* Harper, New York.

Piaget, J. (2013 [1926]). *La représentation du monde chez l'enfant.* 2nd edition. Presses Universitaires de France, Paris.

Pierron, D., Razafindrazaka, H., Pagani, L., Ricaut, F.-X., Antao, T., Capredon, M., et al. (2014). Genome-wide evidence of Austronesian-Bantu admixture and cultural reversion in a hunter-gatherer group of Madagascar. *Proceedings of the National Academy of Sciences of the USA*, 111, 936–941.

Piketty, T. (2013). *Le capital au XXIe siècle.* Editions du Seuil, Paris.

Proudhon, P.-J. (2002 [1846]). *Système des contradictions économiques ou philosophie de la misère*. eBooksLib.com.

Pyle, R. M. (1993). *The thunder tree: lessons from an urban wildland*. Houghton Mifflin, Boston, MA.

Rabhi, P. (2010). *Vers la sobriété heureuse*. Babel, Actes Sud, Arles.

Rand, D. G., Greene, J. D., & Nowak, M. A. (2012). Spontaneous giving and calculated greed. *Nature*, 489, 427-430.

Rifkin, J. (2009). *The empathic civilization: the race to global consciousness in a world of crisis*. Jeremy P. Tarcher/Penguin, New York.

Ripoll, T. (2022). *Pourquoi détruit-on la planète? Le cerveau humain est-il capable de préserver la terre?* Le Bord de l'Eau, Lormont.

Rolston, H. I. (1988). *Environmental ethics: duties and values in the natural world*. Temple University Press, Philadelphia.

Rosenberg, M. (2003). *Life-enriching education*. PuddleDancer Press, Encinitas, CA.

Rosenberg, M. B. (2015). *Nonviolent Communication: a language of life*. 3rd edition. PuddleDancer Press, Encinitas, CA.

Ruskin, J. (1967 [1860]). Unto this last. In: *Four essays on the first principles of political economy* (ed. Lloyd J. Hubenka). University of Nebraska Press, Lincoln.

Sahlins, M. (2017 [1972]). *Stone age economics*. Routledge, Oxon, UK.

Schmidt, A. (1971). *The concept of nature in Marx*. NLB, Londres.

Schumacher, E. F. (2010). *Small is beautiful: economics as if people mattered*. Harper Perennial, New York.

Schwartz, R. C., & Sweezy, M. (2020). *Internal family systems therapy*. 2nd edition. Guilford Press, New York.

Schweitzer, J., & Notarbartolo-di-Sciara, G. (2009). *Beyond cosmic dice: moral life in a random world*. Jacquie Jordan, Los Angeles, CA.

Serpell, J. (1986). *In the company of animals: a study of human-animal relationships*. Basil Blackwell, London.

Serres, M. (1992). *Le contrat naturel*. Flammarion, Paris.

Servigne, P., & Stevens, R. (2015). *Comment tout peut s'effondrer: petit manuel de collapsologie à l'usage des générations présentes*. Le Seuil, Paris.

Shapin, S., & Schaffer, S. (1985). *Leviathan and the air-pump: Hobbes, Boyle, and the experimental life*. Princeton University Press, Princeton, NJ.

Silvertown, J. (2015). Have ecosystem services been oversold? *Trends in Ecology & Evolution*, 30, 641–648.

Singer, P. (1975). *Animal liberation*. New York Review Books, New York.

Soga, M., & Gaston, K. J. (2016). Extinction of experience: the loss of human–nature interactions. *Frontiers in Ecology and the Environment*, 14, 94–101.

Steadman, D. W. (1995). Prehistoric extinctions of Pacific island birds: biodiversity meets zooarchaeology. *Science*, 267, 1123–1131.

Taylor, P. W. (1981). The ethics of respect for nature. *Environmental Ethics*, 3, 197–218.

Toulmin, S. (1990). *Cosmopolis: the hidden agenda of modernity*. University of Chicago Press, Chicago.

van de Waal, E., Borgeaud, C. & Whiten, A. (2013). Potent social learning and conformity shape a wild primate's foraging decisions. *Science*, 340, 483–485.

Weber, M. (2011 [1904-1905]). *The protestant ethics and the spirit of capitalism*. Oxford University Press, Oxford.

Whitehead, A. N. (1920). *The concept of nature*. Cambridge University Press, Cambridge.

Wilson, E. O. (1984). *Biophilia*. Harvard University Press, Cambridge, MA.

Wilson, E. O. (1992). *The diversity of life*. W. W. Norton, New York.

Wilson, E. O. (1998). *Consilience: the unity of knowledge*. Vintage Books, New York.

Index